SEC

ART AND MONUMENTS

EDILERA

La imagen de la Torre de Hércules ha sido proporcionada por la Comunidad del Convento del Santo Domingo el Real de la Ciudad de Segovia.
La imagen del interior de la Casa-Museo de Antonio Machado se se han obtenido con permiso de la Real Academia de Historia y Arte de San Quirce y del Excmo. Ayto de Segovia.
La imagen del interior de la iglesia del Corpus Christi del Convento de Santa Clara ha sido proporcionada por la Comunidad del Convento del Corpus Christi de la Ciudad de Segovia.
La imagen del interior de las ruinas romanas de las plaza de Guevara se han obentido con permiso de la Comunidad de Propietarios bajo la que se asienta.
Todas las imágenes de los instrumentos de medición romanos han sido proporcionadas por D. Isaac Moreno Gallo, http://www.traianvs.net
Todas las imágenes del interior de la antigua cacera del acueducto en la Plaza Mayor y del desarenador de San Gabriel, Real Casa de Moneda, se han obtenido con el permiso del Excmo. Ayto. de Segovia.
Todas las imágenes del interior de la Casa Mudéjar, se han obtenido con el permiso de la dirección del Hotel Casa Mudéjar.
Todas las imágenes del interior de la Cueva de Santo Domingo del Convento de Santa Cruz la Real se han obtenido con el permiso de la Comunidad del Convento de Santo Domingo el Real de Segovia y de la Orden de Predicadores.
Todas las imágenes del interior de la iglesia de la Vera Cruz se han obtenido con el permiso de la Soberana Orden de Malta.
Todas las imágenes del interior de la iglesias de los Santos Justo y Pastor y de San Millán en Segovia, se han obtenido con el permiso del Obispado de Segovia.
Todas las imágenes del interior de la Real Fábrica de Cristales de San Ildefonso se han obtenido con el permiso de la Fundación-Centro Nacional del Vidrio de San Ildefonso.
Todas las imágenes del interior de la S.I. Catedral de Ntra. Sra. de la Asunción y San Frutos, se han obtenido con el permiso del Ilustrísimo Cabildo de la Catedral de Segovia.
Todas las imágenes del interior del Alcázar de Segovia, se han obtenido con el permiso del Excmo. Sr. Presidente del Patronato del Alcázar de Segovia
Todas las imágenes del interior del Convento de Carmelitas Descalzos se han obtenido con el permiso de Prior del Monasterio.
Todas las imágenes del interior del Convento de San Antonio el Real se han obtenido con el permiso de la Comunidad.
Todas las imágenes del interior del Convento de Santa Cruz la Real se han obtenido con el permiso de la I.E. University.
Todas las imágenes del interior del Convento de Santa Isabel se han obtenido con el permiso de la Comunidad del Convento.
Todas las imágenes del interior del Convento de Santa María de El Parral se han obtenido con el permiso del Prior de la Comunidad.
Todas las imágenes del interior del Museo de Segovia, Museo de Zuloaga y de la Biblioteca Pública, se han obtenido con el permiso de la Junta de Castilla y León.
Todas las imágenes del interior del Museo Rodera-Robles se han obtenido con el permiso del Sr. Patrón Secretario de la Fundación.
Todas las imágenes del interior del Torreón de Lozoya y del Palacio de los Condes de Mansilla, se han obtenido con el permiso de la Fundación Caja Segovia.

Agradecimientos:

por la facilidad para tomar fotografías, y buen trato recibido, por orden alfabético:

Cabildo de la S.I.Catedral de Segovia, Sr Presidente-Deán y personal de la S.I.Catedral.
Comunidad de Padres Carmelitas Descalzos de Segovia y encargado
Comunidad de Propietarios de la Plaza de Guevara y Sr. Gerente del Restaurante la Péntola.
Comunidad del Convento de San Antonio el Real y encargada.
Comunidad del Convento de Santa María de El Parral y encargado.
Comunidad del Convento de Santo Domingo el Real de Segovia y Orden de Predicadores.
Comunidad del Convento del Convento de Santa Isabel de Segovia.
Comunidad del Convento del Corpus Christi de Segovia.
D. Isaac Moreno Gallo. www.traianvs.net
Encargado de la iglesia de los Santos Justo y Pastor.
Excmo.Ayto. de Segovia, Turismo; técnicos y guías de la Ciudad de Segovia.
Fundación Caja Segovia y encargados.
Fundación Centro Nacional del Vidrio, Real Fábrica de San Ildefonso, y personal.
Hotel Casa Mudejar, Sr. Director y encargados.
I. E. University y encargados.
Junta de Castilla y León, Cultura, Segovia.
Museo de Segovia, Sr. Director y encargados.
Museo Rodera-Robles, Sr. Patrón secretario y encargado.
Obispado de Segovia, Delegación de Cultura.
Parroquia de San Millán.
Patronato del Alcázar de Segovia, Sr. Presidente, Gerente y demás personal.
Real Academia de Historia y Arte de San Quirce, Segovia.
Soberana Orden de Malta, Sr. Sacerdote de la Vera Cruz y encargado.

Traducidos al inglés por: Simon Harris conforme a la Norma Europea DIN EN 15038

EDILERA
www.edilera.com
contacto@edilera.com

ISBN: 978-84-942521-1-2
Depósito Legal: ZA-78-2014

Impreso en España

TABLE OF CONTENTS

HISTORY

The city of Segovia nestles on a high hill surrounded by the rivers Clamores and Eresma which come together at the foot of the Alcázar[56]; this strategic geographic location, combined with the fertility of the fertile plains of its rivers, gave rise to the formation of the first settlements in the area in around 5 B.C. during the Iron Age.

Celtiberian boar. Segovia Museum

In 3 B.C. the Pre-Roman, Celtiberian tribes were present in the site of the city as is borne out by part of the pit found at Casa Mudéjar[38], right in the heart of the Jewish quarter, also the location where a boar was found and where there was a hill-fort. Other evidence in this settlement are the boars found at Plaza de San Martín[32] and which today houses the city Museum[6].

With the Roman conquest and the subsequent Romanisation, Segovia probably attained municipality status. It is from this time that we have the first written testimony including the term Segovia; it is an "as", a copper Roman coin minted in 17 B.C. Although no-one knew about the Roman city except for the enigmatic presence of the Aqueduct[12], numerous vestiges have been discovered recently in different areas of the city which suggest a major population centre, hence justifying the construction of such a great engineering work.

As from Segovia. Segovia Museum

After the Barbarian invasions and the subsequent Moslem invasion, the city entered a period about which little is known; it is believed that there was some type of settlement which could have raised some large religious construction or other such as the former St. John of the Knights[88] building.

Until the second half of the 11th century there is no documentary record of Segovia; it is known that

Roman ruins under Guevara square

during the short reign of Sancho II the city was set upon by the king of the Toledo "taifa" (Moslem kingdom) and that the latter laid waste part of the aqueduct, cutting off its water supply.

In 1088 the city was repopulated by Alfonso VI through his son-in-law Raymond of Burgundy; from this time onwards Segovia took off as a major city which was to be extremely important in the history of Castile. Its high point arrived when the Trastámara dynasty came to the throne which showed a special affection for the city through Juan II and Enrique IV[102] who spent long spells there; during the reign of the latter Segovia became the backdrop for numerous confrontations between the monarchy and the nobility. In December 1474 Segovia bore testimony to the proclamation of Queen Isabel I, the Catholic[39], at the former church of St. Michael[39]. However, it was to play its most important role during the Guerra de las Comunidades (Revolt of the Comuneros) when it was one of the main cities to rise up against the abusive power that Carlos I wished to impose to the benefit of his Flemish courtesans and shunting aside his mother Queen Juana I; the Segovian Juan Bravo was one of the ringleaders and after the defeat at Villalar he was executed and his property was confiscated which was also the fate of the other participants. One consequence of the revolt was the transferring of the former cathedral to its current location separate from the Alcázar. During the 16th century the city underwent an age of splendour as an upshot of the prospering textile industry, suffering the same fate as the latter when it ceased to be the main economic engine of the kingdom. To add to this industrial crisis there was an outbreak of plague at the end of the century which reduced the population considerably as around 12,000 people died. From this time onwards Segovia sank into decadence along with the rest of the Castilian cities.

In 1985 Segovia was declared a World Heritage City which has boosted its economy.

Statue of Juan Bravo at Medina square

FIESTA DE LA CATORCENA

This celebration dates back to 1410 when the sexton of the now extinct church of St. Facundo went to a Jewish lender to sort out some economic problems, with the latter requiring a consecrated Host in exchange; once the Jew had the Host in his possession he met up with his fellow brethren in the Synagogue to boil the Blessed Sacrament which was miraculously saved as there was a earth tremor and it seeped out through one of the cracks that had appeared in the walls of the building. By dint of this occurrence the king donated the synagogue to the Bishopric and it was thereafter known as the Corpus Christi Church[35]. It was here that this festival in exaltation of the Eucharist was celebrated called the "Catorcena" in honour of the 14 parish churches the city had at that time and who were responsible, one each year, for carrying out the preparations for the festival on the first Sunday in September.

Church of Corpus Christi
Photo: courtesy of the "Hermanas Clarisas" of Corpus Christi in Segovia

SGRAFFITO

When walking around Segovia we can see that the vast majority of its façades are endowed with curious decoration based on elaborate geometric compositions which, like a tapestry, cover the whole wall. This technique, involving the scratching of the plaster coats, is very old and was already used as far back as the 16th century by the Mudejars, having been recovered in the mid-19th century to embellish and lend dignity to the poor-looking façades and afford a different image of the city which had sunk into nineteenth-century decadence. In Segovia we can find more than 300 different sgraffito models.

Four types of sgraffito

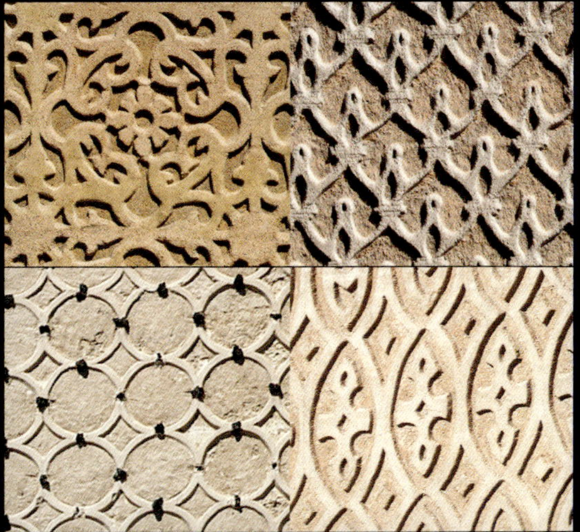

Former Jewish district, Valley of the Clamores, the Jewish cemetery and the Casa del Sol

This institution was founded in **1842** and, after moving around different buildings of the city, was set up at **Casa del Sol**, the former Jewish slaughter-house in the **15th century** situated at the very limits of the "aljama" (Jewish quarter) and belonging to the city defences. The museum houses a collection of varying origins adding up to around 1500 pieces; one of its most noteworthy collections of the excellent coin collection containing coins from the **"ceca de Segovia"** (Segovia Mint), accompanied

Roman mosaic MSG

Visigoth trousseau MSG

by scale models of the machinery used for its production.

A very important section is that dedicated to **Visigoth metal craftwork** with numerous pieces from different necropolises of the province. The visit

Doubloon of the Catholic Monarchs minted in Segovia MSG

to this museum is a tour around the city's history in which we can see everything from prehistoric utensils and Celtiberian boars to pieces of Roman origin like the mosaic brought from the Casa Mudejar[38] or the pieces deriving from the important archaeological site of Duratón[155], highlighting the delicate section of aqueduct and its construction. Self-evidently, a fine collection of **glass** from the **Royal Site of San Ildefonso**[144] has been collected and an area devoted to the textile industry which was so important to Segovia, contributing scale models and illustrative panels; as far as this industry is concerned, special mention must be made of the tableaux by Castilian and Flemish painters from the 15th and 16th centuries carried out right at the height of the industry which allowed the exchanging of works of art with other countries. Its most important works include the retable of the extinct church of Santa Columba, located alongside the Aqueduct[12], the **engravings** by **Durero** and **Rembrandt** and works by contemporary artists like **Madrazo** or **Sorolla**.

El Segoviano, J. Sorolla. MSG

CITY WALLS

The city of Segovia was set up at a strategic place on a rocky scarp surrounded by the valleys of the Rivers Eresma and Clamores; as well as this prime natural position, at the end of the **11th century** and the early **12th century**, after the reconquest by Alfonso VI and the resettlement by Raymond of Burgundy, its city walls were constructed. There has been much speculation about the prior existence of some type of defensive system; hence, in the Alcázar[56] zone and the area surrounding St. Andrew's gate, the appearance of reused Roman material, tombstones and engravings, would suggest the existence of a possible Late Imperial Roman wall, of which the layout in the construction of the medieval city walls has been preserved. Some people also say that this would suggest the existence of an Islamic city wall, but it has not been possible to confirm this to date.

Roman tombstone reused on the city wall: PVBLICIO
IVENAL F
IVVENALIS
Publicius Juvenal, Juvenal's son

The Segovia city wall was declared a **Historic-Artistic Monument** in 1941 and, along with Ávila and Lugo, it is the only Spanish city which retains its **walled enclosure intact**, except for some gates and doors and shutters which, over the course of time, have gradually disappeared for the sake of modernity as is the case of the gates of San Martín and St. John, demolished during the 19th century for urbanistic reasons.

This is a walled enclosure endowed with large dimensions with a perimeter exceeding **3000 metres**, carried out in masonry, brick and limestone and granite ashlar; it once had **13 gates and shutters** of which the following have been preserved: Gates of St. Andrew's, James and St. Cebrián and shutters of Consuelo, St. John, Sol and Rastro, and Luna or the Judería (Jewish quarter), the latter two being highly restored and altered. In addition, the defensive system was strengthened by **fortified houses** which served as support to the gates as is the case of Casa de los Picos[20] at the old Gate of San Martín or Casa de las Cadenas or of Segovia[94] and the Marquis of Lozoya[94] at the now extinct Gate of St. John.

1 Alcázar
2 St.James' Gate
3 San Pedro de los Picos
4 St. Cebrián's gate
5 Marquis of Lozoya fortified house
6 St.Jonh's Gate, extinct
7 Cadenas fortified house
8 Shutter of Consuelo
9 St. Martins' Gate, extinct
10 Picos fortified house
11 Gate of Luna
12 Gate of Sol
13 Roman tombstone
14 St. Andrew's gate or del Socorro
15 Sol fortified house

ST.ANDREW'S GATE[29]

Also known by the name of Arco del Socorro, it opens out into the southern area of the enclosure; it was the main access to the Segovian **aljama**[35] as it connected it to the Jewish cemetery located on the other bank of the River Clamores. During the **15th and 16th century** it was totally overhauled and its current appearance dates back to this time. Two towers, one polygonal and the other quadrangular go to make up the gate space, today taken up by the **City Wall Information Centre**, from which the rampart can be accessed. The inner face houses an image of the Virgin of Succour from where it takes its nickname. In the vicinity a **Roman tombstone** is encrusted in the front of the city Wall as well as a plate recalling the adventures of ***El Buscón***, Mr.Pablos de **Francisco de Quevedo** in the Jewish district of Segovia.

Puerta de Santiago

St.Cebrián gate

ST.JAMES' GATE[35]

Its current name derives from the now extinct Romanesque church of St.James which as situated outside the wall at a site nearby; however, in the 12th century it was called Rodrigo Ordóñez. It is located in the northern area of the city wall and it provides services to the suburb of San Marcos, being crossed every day by **St. John of the Cross**[139] during his stay in Segovia. It was originally a knight's tower, but over time it was joined to the city wall; during the 19th century it was endeavoured to demolish it on two occasions, but it ended up standing the test of time. During the course of its history it has served various purposes, including that of a shelter for the poor and, subsequently, in the 20th century, a house for artists, the first of whom was the painter Celestino Santos Sanz. Its exterior 13th century façade boasts a horseshoe arch, whilst the interior, carried out in the 17th century, has an arch with stone padding. In its interior it has murals with the representation of St. Michael and the city coat-of-arms.

ST. CEBRIÁN GATE[37]

Situated on the northern face of the city wall between the extinct St. John's gate and that of St.James the Eresma valley opens up; the gate we can now see comprises the works carried out in the **17th and 18th centuries** which took the place of the former medieval gate.

THE ACQUEDUCT 1

This major Roman civil engineering work is believed to have been built when Trajan was the emperor. Its dating is backed up by the archaeological vestiges found at pile no.115; there, in the foundations between the excavation filling, there appeared a Trajan sestertius minted during his 6th consulate, in other words, between 112 and 117 A.D., the final years of his imperium, which is why, bearing in mind that these foundations were filled whilst the first rows of stone protruded from the site to facilitate the construction, said dating seems feasible.

Furthermore, there is also the possibility of the reconstruction of the two inscriptions of the attic thanks to a study carried out by the historian G. Alföldy in 1992 based on the marks left on the ashlar by the pins of the letters which were supported with plumb:

IMP•NERVAE•TRAIANI•CAES•AVG•GERM•P•M•TR.P•II•COS•II•PATRIS•IVSSV
P•MVMMIVS•MVMMIANVS•ET•P•FABIVS•TAVRVS•IIVRI•MVNIC•FL•SEGOVIENSIVM•AQVAM•
RESTITVERVNT

By mandate of Emperor Nerva Trajan Caesar Augustus Germanicus, Pontifex Maximus, in his second Tribunicia Potestas, Consul for the second time, Father of the Nation.

Publius Mummius Mummianus and Publius Fabius Taurus, Duumviri of Flavio Council of Segovia, repaired the aqueduct.

The mention of the 2nd consulate of Marcus Ulpius Trajan therein in 98 A.D. seems to contradict the dating of the sestertius in his 6th consulate and this may be explained as follows: construction was started on the mountainside and it progressed to the City to supply it with water as soon as possible without having finished the arches; as this was pillar nº 115, in other words, one of those closest to the City, this theory seems to confirm, after a decade undergoing works, that the inscription would only date the Trajan mandate period and the sestertius the approximate termination date which would not contradict the archaeological evidence.

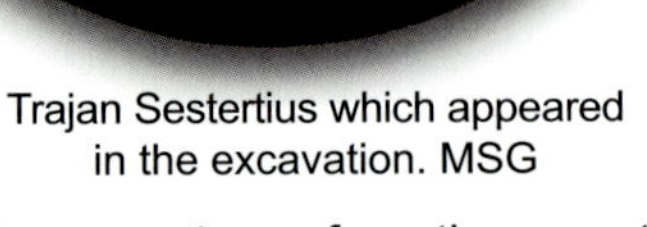
Trajan Sestertius which appeared in the excavation. MSG

What is more problematic is the word RESTITVERVNT because there is no evidence that another aqueduct existed beforehand, although as has been set out by some academics, this could be due to a Damnatio Memoriae against Domitian, with it having been the latter who started the works and Trajan could have taken possession thereof, justifying the restituervnt.

During the Early Middle Ages its history is totally unknown, contrary to the evidence of Visigoth remains in the City which allows it to be assumed that it remained in use. Darkness followed the Moslem invasion until, according to the chronicles, in 1071 the King of the Toledan taifa, al-Mamún, destroyed 36 arches of the lowest part and near the sand remover in a plundering raid designed to cut off the water from the besieged.

In 1088 Raymond of Burgundy, on the orders of Alfonso VI, repopulated the City and there was news that the aqueduct was up and running again which is why it is assumed that the destroyed part had been partially repaired, perhaps using wooden scaffolding. At that time it was already called the «dry bridge» in reference to the fact that no water ran under its arches.

In 1483 the Catholic Monarchs appointed Friar Pedro de Mesa as the person responsible for drawing up the repair project, the prior of the Hieronymus monastery of El Parral, with Juan de Escovedo taking charge of the technical management, also a monk; the reconstructed arches date back to this time from new and old ashlar, some of which have a pointing shape. The irrigation channel (canal) was also cleaned and the waterwheel from where the water came from the mountain.

In 1868, as is borne out by a plate, various arches were restored again in Calle Almira, though using mortar and finishes different from the original.

Since it was put up, the aqueduct has supplied Segovia with water almost uninterruptedly; in 1907 the piping was directed to the pipeline and the canal ceased to fulfil its purpose.

PULL-DOWN OF AQUEDUCT ON CENTRAL PAGES

Reconstruction of a functional chorobates by Isaac Moreno Gallo, the tube is the libra aquaria. Photo: I. Moreno Gallo

TOPOGRAPHIC INSTRUMENTS USED

It should be pointed out that the main success of the Romans with the aqueducts was their capacity to ascertain levels with great accuracy thanks to the chorobates which allowed them to be built with unevenness of up to 0.4 % as is the case of Segovia. With this in mind, the following instruments were used:

Libra Aquaria.

This is a consistent water level in a tube, currently made of casing, filled with water which is used in distances of less than 10 m and allows levelling with great accuracy. In this case the chorobates.

The chorobates.

According to Vitruvius and the fully functional reconstruction carried out by the technical Public Works' engineer Isaac Moreno Gallo of the Public Works' Ministry regarding the Vitruvian description consists of a 20 feet ruler (5.92 m) which can turn on an axis coupled with a base levellable by means of screws and which rests on a tripod. At the end two brackets endowed with plummet levels allow the chorobates to be levelled properly. To calibrate it, a libra aquaria would be used which would allow the final position of the lead wires to be marked, being recalibrated as many times as was necessary. In levelling tests compared with a modern optical level there was more than sufficient precision to build works similar to the Roman ones provided that the topographer had good vision.

Reconstruction of a working dioptra by Isaac Moreno Gallo Photo: I. Moreno Gallo

Dioptra.

According to Heron from Alexandria, Hipparchus and Polybius, and the functional reconstruction carried out by Isaac Moreno Gallo, this is the predecessor of Theodolite. It is one graduated horizontal circle and one vertical one with a graduated semicircle. The horizontal rests on a base with three graduable screws which is levelled by means of a plummet. To measure, in the reconstruction it was decided to opt for a canute which supports the pinholes as described by Polybius in the siege of New Carthage. The completed instrument served for carrying out minor levelling, field measurements, angles, areas, distances and heights.

Recreation of a Roman onsite layout with Dioptra
Photo: I. Moreno Gallo

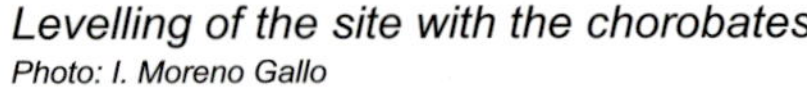

Levelling of the site with the chorobates
Photo: I. Moreno Gallo

CONSTRUCTION

Dam and inlet at an altitude of 1256 m of the crystal clear waters of the River Frío.

Settlement tank at the inlet to eliminate impurities

Before carrying out construction, a design was drawn up by a military or civil engineer who would take charge of the elevations existing between the City and the source, in this case River Frío at the locality of Acebeda.

Once the design had been approved, the layout would be drawn up using a dioptra (a theodolite); immediately afterwards, prelevelling would be carried out using another instrument, the chorobates (a level), and if the slope was exceeded, the layout would be corrected.

Having established the final layout, final levelling would be carried out, using the chorobates again

Path and River Frío

Former irrigation channel and stone wall originally from the locality of El Borreguil, alongside the boundary pillars of Carlos III.

A cacera originating from the aqueduct on the mountain; alongside, the current irrigation channel of Hontoria at a lower elevation with the appearance offered by the old one on the sections opened. Both follow the site level lines.

EXCURSION TO THE INLET

If you wish to visit the aqueduct inlet you must follow the path traced out by milestones leaving from the parking on the back of the reservoir Puente Alta in Revenga.

From there, following the milestones such as the one in the image, after passing over the ditches of Revenga and Hontoria and walking along the site level lines you reach the dam and the settling tank where the water is taken from River Frío or Acebeda. The path is not hard, but its best to take suitable footwear, if possible waterproof. In total, the return journey is around 5 km.

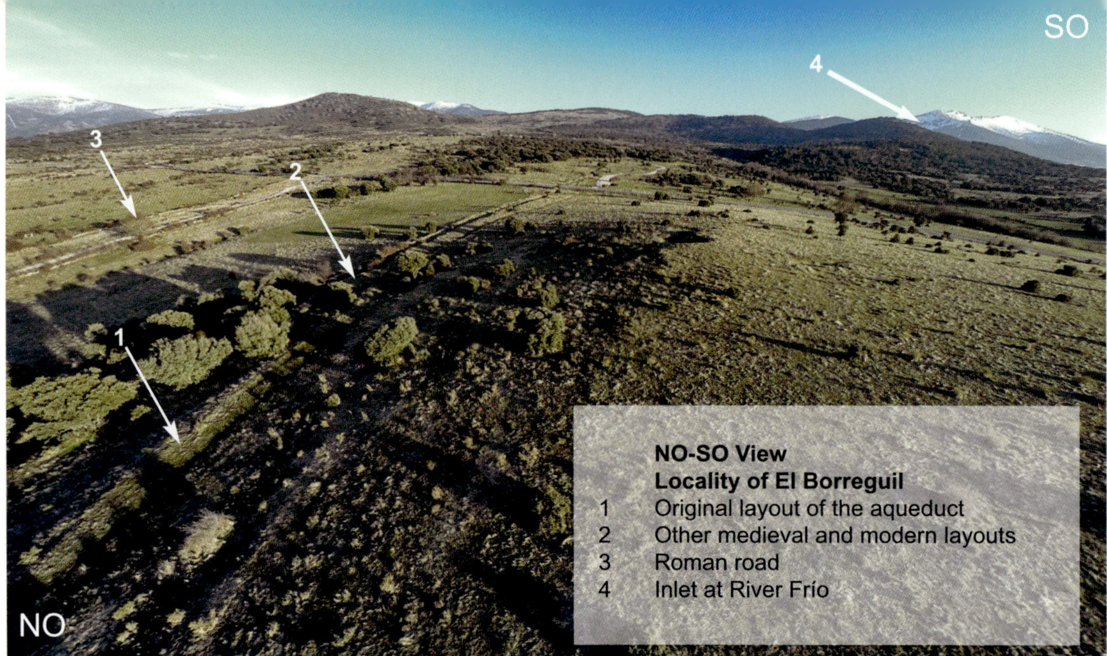

and short sections would be levelled not exceeding 70 m long at a single time and larger sections with the aid of torches and correcting the sphericity of the earth by mathematical procedures.

Finally, at some points of the aqueduct and to guard against any measurement error, if the elevation exceeded that recommended by too much for the appropriate transporting of water and to avoid erosion, gaps would be introduced in the form of manholes and *castella aquarvm* which would allow height and load to be lost without damaging the pipe.

Unfinished canal ashlar at the Roman quarry of El Berrocal at Ortigosa del Monte

At the construction stage, at the intake on River Frío a dam with a spillway was put up and a bypass channel where the water, by means of lime pools, sedimented the solids in suspension and was oxygenated by means of small waterfalls.

The Roman *specus*, the

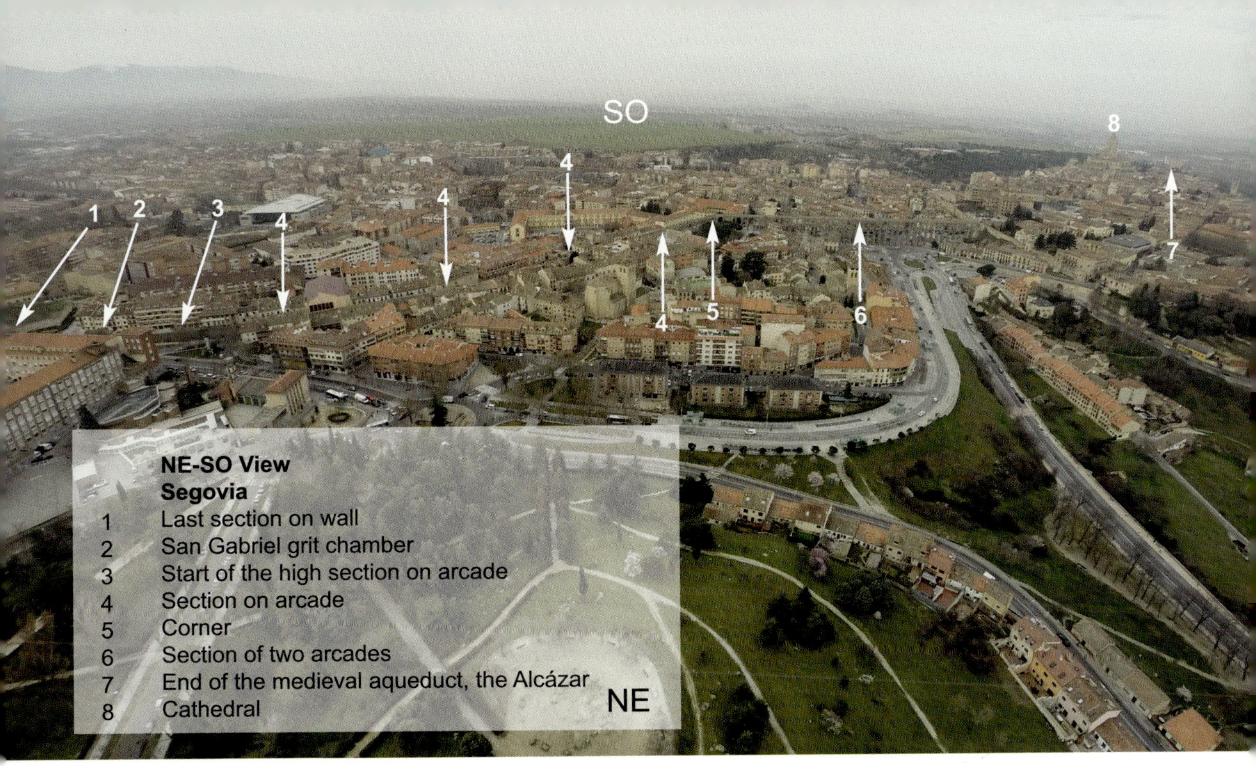
SO
NE
1
2
3
4
5
6
7
8
NE-SO View
Segovia
1 Last section on wall
2 San Gabriel grit chamber
3 Start of the high section on arcade
4 Section on arcade
5 Corner
6 Section of two arcades
7 End of the medieval aqueduct, the Alcázar
8 Cathedral

canal, was nearly all covered until its arrival in the City in order to avoid contamination and sun exposure, maintaining the low temperature of the inlet. The interior was covered by a hydraulic mortar, the *opus signinum*, based on lime, brick and ground tiles which served to waterproof the canal and ensure the cleanliness of the water. During the Middle Ages the specus was replaced with an irrigation ditch which was mainly uncovered.

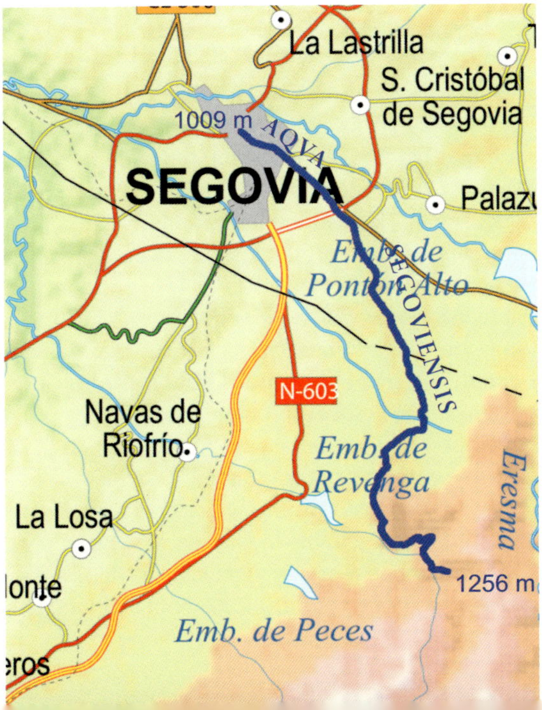
La Lastrilla
S. Cristóbal de Segovia
1009 m
AQVA
SEGOVIA
Emb. de Pontón Alto
SEGOVIENSIS
N-603
Navas de Riofrío
Emb. de Revenga
Eresma
La Losa
1256 m
Emb. de Peces

On its arrival at the City two grit chambers were inserted again, like the current ones whose factories produced large Roman ashlar blocks can still be seen. They consist of a lime pool with an inlet and an outlet opposite on the upper part and a bottom outlet for cleaning the silt and other sediments.

Finally, what is now called an aqueduct started which is a small part of the whole of Roman *Aqua*; it is a 767 m long bridge with 123 arches, 44 of which are double-height. Its construction is similar

Dressed ashlar with one pile. The chiselling (1) can be observed here to provide it with its dressed shape, the hole for inserting the ferrei forceps (2) and the vectis (3)for inserting the lever which to settle it.

1
2
3

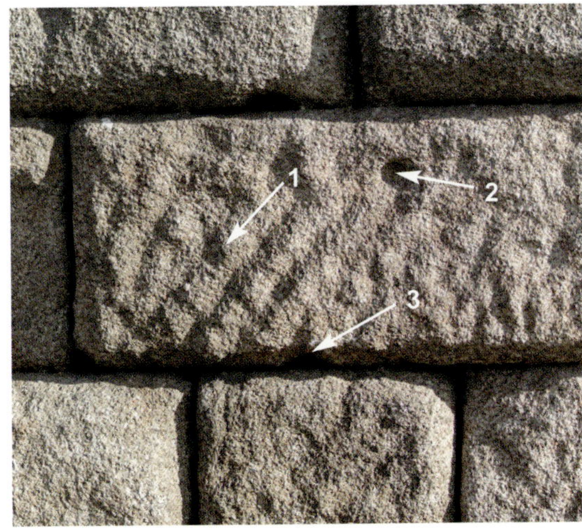

San Gabriel grit chamber

to that of a Roman bridge with a series of piles settled on the rock in *ad hoc* foundations that were later filled with the remains of the final carving and the excavation earth. The piles gradually get smaller in section as they rise. The ashlar of the whole work is dressed and it is settled dry without any mortar, alternatively having ropes and logs to hold it firm.

The ashlar was raised by means of cranes that would take them by the holes which are still visible on their faces and where the ferrei forceps or raising clamps would be introduced. They were then pushed into their final position using levers

Interior of the San Gabriel grit chamber; via the left downwards, water inlet, sedimentation in the lime pool and outlet in front; excess flows via the left and silt via the bottom.

which would be introduced into the vectis, or grooves, which appear on the upper vertices.

As a result of having been settled dry, the aqueduct enjoys a certain flexibility with regard to seismic movements, the broad daily thermal oscillation and other own and external actions. The movement of the aqueduct over time gradually separated the two leaves of the upper arches, at least since before the 15th century.

The aqueduct itself runs on the crown which, in line with the archaeology, seems to date back to the restoration of the Catholic Monarchs.

Finally, having passed over the city walls, there would be a *castellum aquae* where the canal would

Interior of the canal in the Plaza Mayor record around the 15th century; on the right, a diversion or merced can be seen on the canal wall.

start to be spread around the City, prioritising the supply to public fountains, cisterns and baths such as those discovered on the south atrium of San Martín in 1864 and finally those which had a water concession. The distribution would be carried out by means of underground canals, overpasses and ceramic and lead piping. The current one reaches the Alcázar does not seem to date back any further than the 15th century and hence the original Roman distribution is not known, though it is believed that in Queen Juana square there must have been a *castellum aquae* for distribution.

The diversions or Mercedes.

From the construction the water was supplied to the fountains and to the mercedes. The latter were public water concessions to private individuals who requested them in return for a contribution. The

Corner of the aqueduct at Plaza Díaz Sanz; here the aqueduct passes from one arcade to two

Image of the Virgin of Paz in the niche of the aqueduct console. Cadets from the Artillery Academy go up to it on the day of their patron saint, St. Bárbara, to decorate her with the Spanish flag.

public fountains always took priority over any other uses.

The free pipe system, used until the installation of meters, consisted of letting the water run freely at all the water exits with a view to ensuring that the water was always in motion, avoiding any accumulations of pathogens and any dirty in addition to the cleaning of the sewers with excess water from the fountains and mercedes. The majority of the excess water would usually end up in the Clamores stream bordering the walled City to the west and which stemmed from the remains of the lime pools and the excess from the first mercedes of the aqueduct.

For the mercedes concession a hole would be carried out in the canal with regard to a measure submitted to the Council and which was usually coin. There was fraud and also illegal inlets such as the cerbatanas which would steal water by means of a siphon and by suction; other inlets were legal even though they used this same system.

GENERAL DATA ABOUT AQUEDUCT

- Length around 15 km.
- Length on arcades: 767 m
- Minimum slope: 0,4 %
- Mean slope: 1,64%
- Maximum transportable flow in a worst case scenario a 2-foot canal (approx. 0,6 m): 5,184 m3/day

CASA DE LOS PICOS[2]

This is undoubtedly one of the main monuments in Segovia owing to its unique façade covered rhythmically and almost completely, as if it was a tapestry, by huge **diamond tips** which lend it a spectacular set of volumes and chiarascuro. This house belonged to López de Ayala y Silva who later sold it to Alonso González de la Hoz whose coat-of-arms hangs above the doorway. Theories abound about its

Detail of the façade of the house of Picos

main decorative element, as occurs with the contemporary Casa de las Conchas de Salamanca; some people argue that its appearance can be put down to its defensive nature as it rises alongside the extinct San Martín gate; it also alludes to the Italian-based influence characteristic of the late 15th century, the time when the façade was built. What's more, we can find another more popular explanation which refers to the property having belonged to a Jew or an executioner and that the new owner, to erase the memory of this past, deployed this decorative element to thereby change its denomination.
The building is currently taken up by the School of Art.

Façade of the house of Picos

CASA DE LOS DEL RÍO[3]

This building, which usually goes unnoticed to the onlooker owing to its simple **Gothic façade** which is decentered with regard to the street, hides inside one of the jewels of the city. The house is old in origin as is borne out by the **Romanesque remains** of an arch opening onto the **courtyard** and the most notable aspect is the latter, comprising seven columns with helicoidal shafts which are unique in the city.

THE CORN EXCHANGE[5]

The former corn warehouse for the Segovian Council is currently taken up by the Municipal Archive which stores important documents linked with the history of the city and also of the kingdom of Castile such as the **Act of Proclamation of Queen Isabel the Catholic** on December 13th 1474.
The building dates back to the early **16th century** and it bears a simple gateway framed by an alfiz decorated with balls and two coats-of-arms of the city.

COUNT ALPUENTE PALACE~HOUSE 4

Gothic span with tracery

This wonderful example of late **15th century** civil architecture was built by Mr. Alonso Cascales and his wife Mrs.Ana de Barros, transferred in the 19th century to Francisco Javier Aspiroz, the 1st Count of Alpuente, which is why it is also known as the House of Aspiroz.

Although the building was constructed in the 15th century, it seems to have been raised on the remains of a house from the former Moorish quarter situated in this area as is borne out by a horseshoe arch discovered during a restoration alongside the main gateway. Three **windows** are opened out on the latter with rich Gothic tracery and fine carvings on the chapters that support them. Inside the interior building there lies the **courtyard**, today covered by a glazed ceiling decorated by coats-of-arms on the chapters.

MIRADOR DE LA CANALEJA[6] C/ CERVANTES

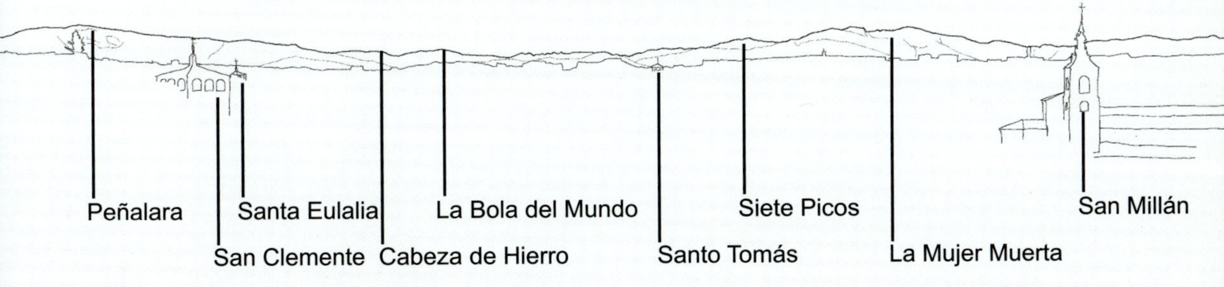

Peñalara

Siete Picos

LEGEND OF THE DEAD WOMAN

Many legends have arisen around the strange profile that appears in this area of the Guadarrama Mountains. The most widespread recounts the story about the wife of head of the tribe that lived in Segovia who had two twins who, when they grew up, confronted each other for the sake of power; to avoid a terrible outcome, the mother gave her life up to the Gods in exchange for a truce between the brothers which is why at the time of the fight a blizzard rose up which covered everything in snow, with the silhouette of the mother appearing on the horizon in a place where there had been no mountain before; tradition has it that every day when it gets dark two small clouds come close to the mountain which represent the twins kissing their mother. Another of the legends which is more widely accepted is one which refers to the fight between two men over the same damsel who, whilst they were fighting, put herself between them and ended up dead, after which there was a great earthquake which gave rise to the formation of a mountain in the shape of the woman. Tradition has also linked it to the figure of Hercules, the legendary founder of Segovia who, at the behest of the king, carved out on the rock the figure of his dead daughter.

La Mujer Muerta

PLAZA DE MEDINA DEL CAMPO

St.Martin's Church and statue of the comunero Juan Bravo.
The square was named after him in honour of the town of the same name which, during the Guerra de las Comunidades (Revolt of the Comuneros) against Carlos I in 1520 preferred to succumb to the flames rather than give up the artillery pieces to lay siege to Segovia.

ROYAL PALACE OF ENRIQUE IV[11]

Mullioned window on Calle Arias Dávila

Also known as the **Palace of St.Martin**, it took up all of the current block and was made up of various randomly arranged constructions, the oldest known of which dates back to the Romanesque period as borne out by two arches found during the rehabilitation of part of the whole to build the Esteban Vicente Museum. In around 1455 the houses were remodelled, distributing the rooms of the king and queen with a Mudejar-style distribution, very much to the taste of Enrique IV around patios and with richly decorated rooms.
In 1510 the palace ceased to belong to the crown and ended up being divided up between various Segovian lineages and any remains of the former palatial unit gradually faded away, retaining the original elements at the **Palace of Queen Juana** or "Mercado Peñalosa" house with a 16th century façade adorned by the coats-of-arms of the new owners; the building is in a terrible state of repair and part of the Mudejar plasterwork decoration on the courtyard remains from the 15th century.
The **King's rooms** are occupied by a 16th century construction known as **Mexía Tovar house** whose coats-of-arms on the façade are attributed to the master builder of the cathedral **Pedro de Brizuela**. On one side some original mullioned windows from the 15th century can still be seen, the sole testimony of the palace that existed there.
The central part which joined both palaces is now taken up by the Municipal Market as well as other buildings and the Contemporary Art Museum; after the division, this area was occupied by the Hospital de Nuestra Señora de la Concepción, also known as the Hospital de Viejos (Old

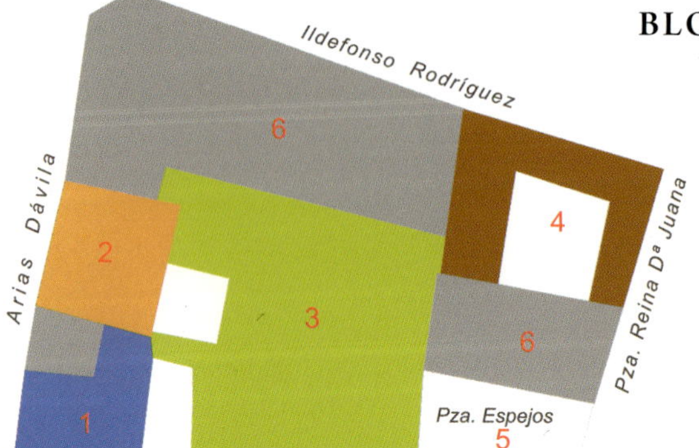

BLOCK OF THE ENRIQUE IV PALACE

1 *The King's Room ~ Casa de los Mexía-Tovar.*
2 *Remains from the 15th century, mullioned windows.*
3 *Esteban Vicente Museum, union between royal rooms.*
4 *Queen's Rooms ~ Queen Juana Palace ~ Casa de los Mercado Peñalosa.*
5 *Plaza de los Espejos, the former Leon's Den of Enrique IV.*
6 *Other modern buildings on the former palace.*

Esteban Vicente Museum, rest of the palace

Gateway to the Mexía-Tovar palace

People's Hospital) as it was founded to house old people without any means of sustenance; over time it fell into disuse and became an Art School. Today, the hospital chapel, along with recently built outbuildings, houses the art collection of the Segovian painter **Esteban Vicente**, a representative of Abstract Expressionism who developed his career in the United States at the Nueva York School, comprising great 20th century artists like Rothko, Pollock, De Kooning and Kline. Esteban Vicente's work is characterised by having converted the abstraction into a means of expression that accumulates feeling and reflection, seeking, first and foremost, to be expression, though guided by a compositional order both in terms of the formal and the chromatic; they are works which are open and dynamic with a tension which could be interpreted as a musical impression in which the colours vibrate at different speeds, marking the intensity and meaning of the painting, a quasi-silence but not silent.

Entrance to the Esteban Vicente Museum

15TH CENTURY HOUSE 7

This house is located in the environs of Plaza de San Martín and it complements the civil architecture group where it is set. For a long time this house was named as **Juan Bravo House** but it wasn't owned by the comunero leader who had his abode not too far away and which as submitted to demolitions and decayed after the Comunidades in retribution for the insurrection. The building we can see was built in the late **15th century**, though there is documentary evidence that the high gallery was added in the mid-16th century; its façade boats elements which are characteristic of the Gothic such as the gateway framed by a broken alfiz adorned by diamond tips or the parapets with Gothic outlines and the ogee arches of the gallery, with the latter having Renaissance roots, binding with the Italian urban residences and which in Segovia very frequently appear in the houses of the time, complying with a specific purpose: the drying of the wool materials from the flourishing textile industry.

COUNT OF BORNOS HOUSE 9

Also known as the **"Mayorazgo de los Galachos" house**, this **16th century** construction has belonged over time to various Segovian lineages until finally coming into the possession of Viscount of Altamira de Vivero.

Situated in an area where noble palaces and houses abound, this is a house which stands out for its simplicity, highlighting, on its façade, the double-glazed corner window characteristic of the 16th century and its façade whose gateway is flanked by two Renaissance-style busts and rounded off by a balcony framed by semicolumns which support a triangular front sheltered by the coats-of-arms of the Peralta, the former owners of the house. The building is topped off by the characteristic Segovian gallery.

CASA DE LOS SOLIER[10]

A model of the Renaissance house, it was known as the **Casa de Correos** (Postal House) as it was owned by the Postmaster General of Segovia Mr. Antonio de Figueredo in the early 17th century. It boasts a beautiful façade framed by two overlaid semicolumns finished in sconce; the main axis of said façade is taken up by the overlaying of the gateway and balconies in decreasing order which lends rhythm and harmony to the whole. It is rounded off by the typical Segovian gallery comprising segmental arches.

FORMER ROYAL JAIL[13]

For many centuries the building today housing the Public Library served as the royal jail within whose confines the distinguished writer **Lope de Vega** spent some time. The building was gradually changed over time and **Pedro de Brizuela** was involved therein, the master builder of the Cathedral, corresponding to the end of the **16th century** and the early 17th century, the current outer appearance of the façade with three heights with a gateway crowned by the Bourbon coats-of-arms which replaced that of the Austrias in 1737 and the turrets that adorn the four corners. In 1942, during the remodelling work to adapt the building to its new use as a Library, the remains of the **Romanesque church of San Medel** were brought here, originating from a deserted spot near Bernuy de Porreros; Hence, in the hallway there is what may have been the triumphal arch with capitals adorned by figures the main gateway of the Romanesque building is situated at the access to the reading room as well as a window and another simpler arch, also corresponding to a gateway.

Remains of the Romanesque church of San Medel

LOZOYA TOWER 8

This is the name given to the unit formed by the tower itself and a Renaissance palace attached to it; the tower fits in with the fortress-house model characteristic of the 14th and 15th centuries, being endowed with a defensive, robust appearance which, in the case in point, has been modified to allow access to the palace erected behind it. This is organised around two courtyards whose beauty is quite breath-taking; the first one is porticoed on two of its sides by an architraved structure resting on intricate baseplates on which there are medallions with the busts of characters from Classical Antiquity. The second of the courtyards is gardened and its configuration is reminiscent of Italian loggias; the decorative resources and solutions of the first courtyard are repeated here.

The whole we see today is the upshot of the numerous remodelling works it has endured over time, having changed owners several times with each lending their own identity thereunto, highlighting the Marquis of Lozoya who bought it in the 18th century and whose name it bears.

Vespasianus Medallion

ST. MARTIN'S 12

This Romanesque church is situated in the old town of the city alongside the main road and surrounded by palaces and fortified houses; it is erected on the remains of some old **Roman thermal bathes** and as far as can be gleaned from the different archaeological remains founds, it was the Roman population nucleus. As far as its origin is concerned, there is a theory that talks about the presence of a **Mozarab church** on the same plot and which would be the seed of the current one; this pre-Romanesque church would correspond to the nine segments of the naves.

The church consists of three naves with a transept marked in height and three apses, the smaller, Romanesque sides and the central one endowed with larger dimensions and corresponding to remodelling undertaken in the 17th century. According to said theory, in the 12th century the original place of worship was expanded, adding to the original nucleus the chevet, the transept and the tower. The constant remodelling work and interventions to which the monument has been subjected make it very hard to gain a clear reading of the chronology of its architecture.

On the exterior its gateways and its porticoed galleries stand out; the **western gateway**, the main one, has monumental development similar to the present one at the church of San Juan de los Caballeros[88] with a protruding narthex richly decorated by column-statues which could well represent apostles or prophets; the design and formal appearance of the **column-statues** are reminiscent of

Caryatids of the western portico

St.Martin's from the Torreón de Lozoya; in the background, from left to right the Cathedral, St.Michael and St.Stephen.

those which decorate the southern gateway of the basilica of **San Vicente de Ávila**, the source which inspired the majority of Segovian Romanesque temples, though the latter are more coarsely are larger in size and their position is more static. At the Baroque top another statue has been repositioned endowed with characteristics which are similar to the previous ones which tradition identifies with the patron saint of the Church, San Martín Obispo.

The **southern gallery** has been totally restored and the majority of its arches and capitals have been built recently; in the interior alongside the gateway there are the remains of some mural paintings whose formal characteristics are similar to the first Gothic one which portrays Pantocrator surrounded by Tetramorphs.

The **northern gallery** does not have a better state of repair than the previous one; however, despite the erosion and having remained closed to accommodate the chapels, though rich iconography related with Christ's public life is recognisable in these capitels; this is followed by a series of **capitels** on which scenes from the Annunciation are shown, the Flight to Egypt, the slaughter of the Innocents, the Last Supper or the Arrest of Jesus, along with others of a profane nature such as the fight between some riders or the motions of some dancers.

THE JEWISH QUARTER

During the 13th, 14th and 15th century the city of Segovia had a major Jewish population; bearing testimony from this time is the Jewish quarter which still conserves its former layout of narrow, winding streets to the south of the city and within the walled site. This district is the upshot of various events as at the beginning the Jews lived in any area of the city, but in the early 15th century this situation changed; in 1410, after being accused of profaning a Host inside its main synagogue, the latter was requisitioned and converted into a Christian place of worship and two years later by the influence of Queen Catalina de Lancaster, the **laws of Ayllón** were proclaimed which brought together, amongst other prohibitions and obligations of the Jewish community, that of being confined in a district, with these provisions being resumed at the time of the Catholic Monarchs. This district had eight doors open to the city and five synagogues as well as other elements characteristic of this community such as the old butcher's located at Casa del Sol, the current Provincial Museum or the cemetery located opposite the Jewish quarter on the other bank of the River Clamores.

With the expulsion of the Jews in 1492 the district took on the name of Barrio Nuevo (New District) and to be occupied by Christians, though the passage of time has not disfigured its appearance and it still retains part of its medieval charm. At the entrance to the former Jewish quarter was the main system which converted to Christianity in 1410 at the church of **Corpus Christi**[14]. In 1899 a fire swept through the building and some perimetrals walls and the dividing arcades between the naves were left standing, with the coffered ceilings and the decoration disappearing; Recent restorations have endeavoured to return it to its original appearance, using old photos and engravings. Its structure is reminiscent of the synagogue of **Santa María la Blanca de Toledo** and so its construction can be dated back to the 14th century. The decorative elements worthy of special mention are the pineapple chapters and reliefs and the small horseshoe arches which form the upper gallery.

Old Jewish quarter street

Located right in the heart of the district is the house of Rabbi **Abraham Senneor**, one of the most distinguished members of the community; this figure played a very important role during the reign of the **Catholic Monarchs**; he played an active role in the negotiations regarding the marriage between Isabel and Fernando and during the reign of the latter he held high posts in the Court, including that of Taxmaster General for the Kingdom of Castile. In 1492 when the Catholic Monarchs enacted the decree to expel Jews from Spanish territory the rabbi was of an advanced age and converted to Christianity and his Godparents were the Monarchs themselves, taking the Christian name of Fernando in honour of the King and as his surnames Núñez Coronel, belonging to an extinct noble lineage. Of his descendants it is worth mentioning his granddaughter **María Coronel**, married to **Juan Bravo**, the comunero ringleader from the city of Segovia. Over time the house underwent various transformations and in the 16th century it became the abode of Andrés Laguna, the physician to King Carlos I and was acquired in the early 20th century by the Bishopric to accommodate a Franciscan convent. It is currently occupied by the **Learning Centre for the Jewish quarter**[15], the place where former habits and customs of the district are taught as well as some rituals characteristic of Judaism.

Another of the Jewish places of worship which has been partially conserved is the **synagogue of Los Ibáñez** whose plot is currently taken up by a religious school as it was sold to the Council slightly before the Jewish expulsion in 1492, thereafter belonging to the Ibáñez family in Segovia who it belonged to until the 19th century when it came back into Church hands again. During some remodelling work carried out at the school in the 1980's the remains of the former synagogue were found such as the **Mikveh** or pool for ritual baths and two windows, one of them endowed with similar characteristics to those of the cathedral cloister, carried out by Juan Guas, and another in porthole form with fine plasterwork. By studying these remains it has been revealed to be structurally similar to the **Tránsito de Toledo synagogue**.

LA PLAZA MAYOR

The Plaza Mayor (Main Square) space was devised in the second third of the 16th century after the subsidence of the Romanesque **St.Michael's** Church[39] which took up its centre. Over the centuries the current buildings surrounding it were gradually constructed, highlighting the **Town Hall**[18] designed by Pedro de Brizuela in the early 17th century, following the architectural models of the time. On another side of the square there lies the neoclassical-stylie **Juan Bravo Theatre**[19] inaugurated in the early 20th century whose craftsman was Cabello Dodero who was also responsible for the neoplateresque style **Casa de los Larios** on the arcades of the former palace which close next to the chevet of the Cathedral. In its centre is the **music kiosk** carried out by Joaquín Odriozola in the late 19th century according to iron architecture traditions.

LA CASA MUDÉJAR[16]

On the main street of the old town is the **15th century** building currently occupied by a hotel whose owners wished to restore and conserve it, thereby boosting the wealth of heritage in the city; this action was worth an Honourable Mention in the context of the **Regional Prize AR&PA 2006 for Restoration** which, for the first time, was handed over to an individual. Its walls conceal more than 2000 years of history, affording valuable information about the city from its early days as a Vaccaei hill-fort as the basements house the remains of the former **Pre-Roman pit**, a discovery added to by the appearance in the 20th century of a **boar** which is kept at the National Archaeological Museum. This area was also the site of the appearance of very important **Roman vestiges** such as a large cylindrical base and a mosaic which can be seen at the city Museum[6]; however, the most surprising and important find was undoubtedly some large stairs with a podium made of masonry which accrue to other discoveries from the Roman age in Segovia, bearing testimony to the existence of a city which was sufficiently large as to justify the presence of the aqueduct. In addition to these remains, the most noteworthy aspects of the building are its **Mudejar coffered ceilings** and the structure of the medieval dwelling; the person who lived here must have been in a healthy economic position and have had some sway in the Court in view of the rich ornamentation on show in various rooms. There has been speculation about his identity, traditionally attributing ownership to Rabbi Mayr Melamed, a tax collector and the son-in-law of Abraham Seneor, converted to Christianity in 1492 under the name of Fernán Núñez Coronel; his social position would explain the double entrance to the house with one door open to the Jewish quarter and the other to the Christian city as well as the presence on the coffered ceilings of heraldic coats-of-arms representing a golden castle in gules crowned by a star, an element alluding to a post close to the king, in this case Enrique IV in view of the similarity of the representation of the castle with the coins minted by this monarch as well as by the appearance of pomegranates amongst the vegetal motifs which adorn the panels, an element which, on the one hand, is a symbol of the motto of the kingdom of Enrique IV, but which is also relevant in the Jewish world as it is one of the fruits of the Promised Land and the belief that it has the same number of seeds as the commandments of the Torah. This old 15th century house still has many secrets to reveal and there has been much speculation and many unanswered questions which are recounted to the visitor by means of guided tours that the establishment offers free-of-charge to spread the word about its rich heritage.

Celtic pit

Roman stairs

Medieval door

Coffered ceiling in the dining room

ST. MICHAEL[17]

The current building was erected in the **16th century** after the subsidence in 1532 of the Gothic vaults raised by Juan Guas in the previous Romanesque temple. After its ruin, it was decided to transfer it to its current plot on one side of the square for the purposes of the construction of the new Cathedral[40]. The features of the new church are attributed to **Rodrigo Gil de Hontañón** owing to the style is represents, very similar to his modus operandi; it consists of a nave with ribbed vaults in Hontañón style. Materials from the former Romanesque parish church were reused in its construction such as ashlar stones or the remains of decorative elements such as metopes with flowers or **figurative reliefs**, highlighting those dedicated to St.Michael, St.Peter and St.Paul which are stylistically related with the western gateway statues of the St.Martin's Church[32]; the current ones are replicas of the originals which are kept at the entrance to Plaza Mayor[36]. The Romanesque church was raised in the centre of the current Plaza Mayor, being one of the most important temples in the city not only because of its dimensions, but also in terms of its historic relevance as the municipality would meet up in its environs to take decisions and it was the setting for the **coronation of Queen Isabel the Catholic** in 1474.

ISABEL THE CATHOLIC

The relationship of Queen Isabel I of Castile with **Segovia** was decisive for the access of the latter to the throne. Born in Madrigal de las Altas Torres in 1451, she spent part of her childhood in Arévalo with her brother Alfonso and her sick mother until when she was 10 years old both children were claimed by King Enrique IV, her stepbrother, on the grounds of the birth of his daughter Juana; they were taken to Segovia where the court was situated, thereby controlling them so as to prevent factions opposing the King from using them against him, which was what actually ended up happening. From this time onwards Isabel lived in the city intermittently, staying both at the **Alcázar[56]** and at the **Royal Palace of San Martín[26]**. After the death of her stepbrother and in line with that agreed upon in the Treaty of the Toros de Guisando, she proclaimed herself Queen of Castile on December 13th 1474 at the **St.Michael's Church** in Segovia; the act of her proclamation is kept at the Municipal Archive, the former building of **Alhóndiga[21]**. Her reign had a rocky start owing to the civil war facing Isabel's faction who supported her niece Juana as Queen; her reign was also marked by numerous historic feats such as the conquest of the Kingdom of Granada or the discovery of America, inter alia, which meant there had to be an itinerant Court. However, Isabel returned on several occasions to Segovia, staying at the same places as in her youth, maintaining her relationship with the city until her death in Medina del Campo in 1504.

ST. MARY AND ST.FRUCTUS CATHEDRAL

HISTORY

Segovia cathedral was the last **Gothic** one raised in Spain in the Renaissance, the same as the **new Cathedral in Salamanca** which served as the model to put into effect the resources tested there. Before the current cathedral there was another one from the **Romanesque** era built during the 12th and 13th centuries alongside the alcázar[56]. According to testimonies from the time, it must have been very similar to **Zamora Cathedral**; it consisted of three naves, a tower and cloister and was situated within a conglomerate that included the episcopal palace, a hospital and the priests' district. This whole unit entailed a hazard from the very outset and it was inconvenient to defend the owing to its proximity. On various occasions they wanted to move but this did not occur until the reign of **Carlos I** when, during the **Guerra de las Comunidades** (Revolt of the Comuneros), the comuneros occupied the cathedral entering via a hole opened in the Main Chapel wall; the royalists took the relics of San Frutos and his brothers to the Alcázar chapel and there were intense struggles in the naves until finally the tower was taken, laying siege to the fortress for six months. With the defeat at **Villalar** it survived the demolition of the old cathedral which had been left marked in a poor condition, though not enough to demolish it and his **change of status**.

The current cathedral started being built in **1525** on the plots formerly taken up by the Jewish Aljama and Santa Clara Convent up to the highest area of the city. The outlines of the new temple were given by **Juan Gil de Hontañón** who died the following year. Some elements of the former cathedral were taken there, as well as the Gothic **cloister** or art of bars. The western gateway, which opened out onto Alcázar, known by the name of **Portada del Álamo**, attributed to Juan Guas, was transferred to the parish of Fuentepelayo on 1523 where it still remains.

During the course of the works, the chapel of the Santa Clara convent continued to be used, located in the cathedral chevet and the last area to be erected. It was consecrated on August 15th 1588 on Assumption Day to whom the temple is devoted.

CONSTRUCTION

The Segovia Cathedral building is the upshot of construction in two different stages and for a period of **150 years**, though adjusting to the original outlines provided by **Juan Gil de Hontañón** which brought about a homogeneous unit where the external influences are minimal. The City and its fervour-driven inhabitants towards their new Cathedral helped to lay the foundations, even at night-time in such a way that in 15 days they had already been dug. To make them settled, there were civic processions during which surplus material was brought from the Old Cathedral and was left at the factory. Notwithstanding the current Cathedral representing the commitment of the Council and the Segovia council to a building which was useful and representative, in addition to being an embellishment to the City.

The **first construction** stage ran from 1525 to 1526 and this corresponds to that outlined by Juan Gil de Hontañón who, owing to his death in 1526, would not actually see nor raised the walls; to take his place he called upon his son **Rodrigo Gil de Hontañón**. Thanks to the donations and contributions it was put up quickly; the works were started in order to be able to keep using the church of the former Corpus monastery which was in the present-day chevet. Priority

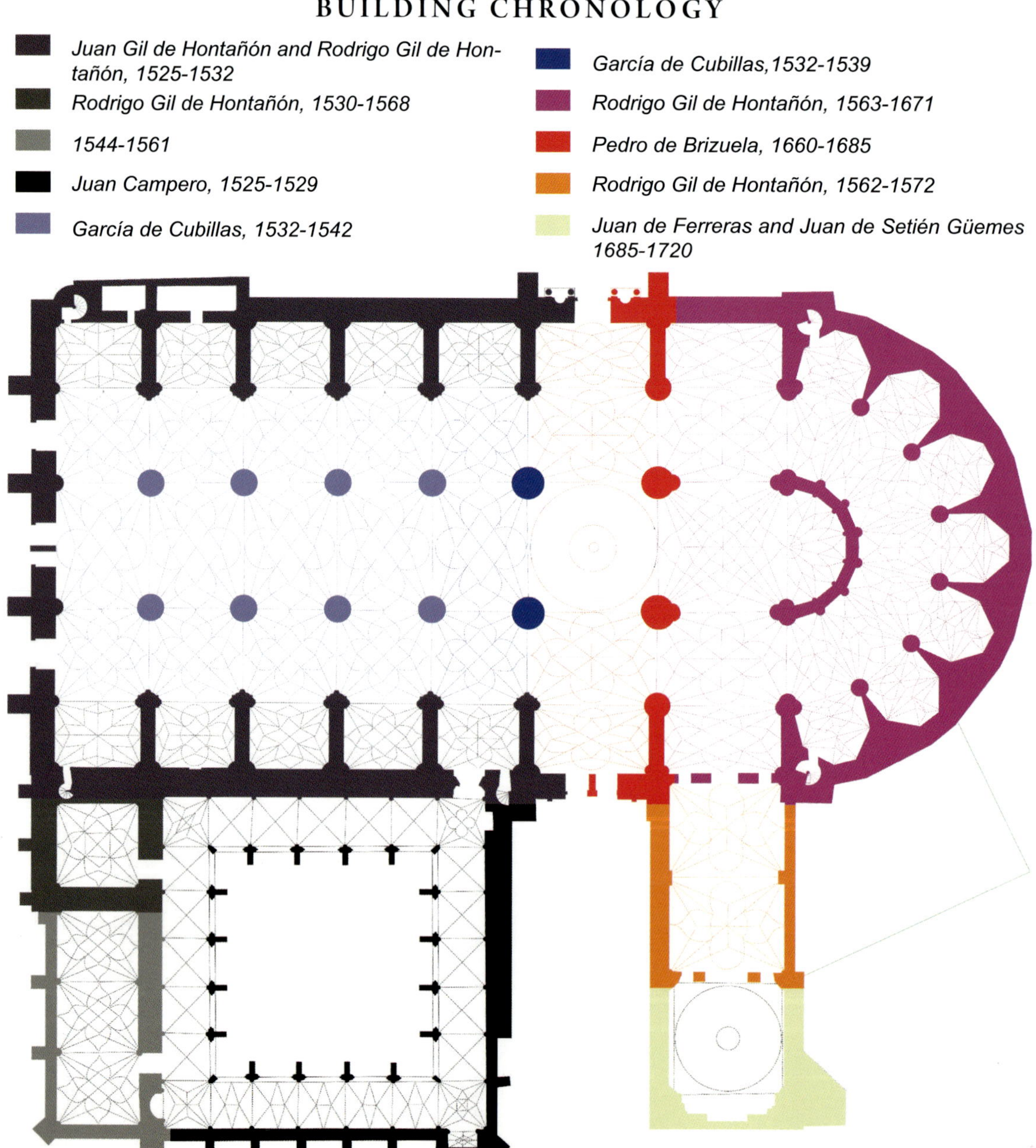

was given to the southside in order to be able to transfer the Gothic cloister of the former Cathedral as soon as possible which Juan Campero had already started dismantling Juan Campero. In 1527 the nave of the epistle was being traced in the sand. In that same year he was dismissed owing to his failure to help out on Rodrigo Gil de Hontañón's works, with his lace being taken by **García de Cubillas** and the churchwarden, **Juan Rodríguez**, who remained true to the original features. On March 5th 1532 Enrique Egas stated that eight of the niche chapels laterals had already been covered from the transept to the base, with the two from the gable still not being covered and that the walls already contained the west gates and the tower was already 60 feet high, becoming one of the highest in Spain. In 1535 the City Council, aware of the beauty and image for the City, asked that nothing be built alongside the Cathedral.

In 1539 all the niche chapels and vaults of the first stretch from the transept were closed. Then the whole of the central nave was covered in 1541 and in 1542 that of the lateral naves. Enrique Egas made a point of stressing that the core of Toral pillars should be made of well-squared masonry and not of concrete rubble of lime and pebble to avoid the subsidence of the domes like in Seville and Burgos. The works continued, renting out scaffolding which looked out onto the Plaza Mayor to see the bulls in the festivals of San Juan, Santiago and Nuestra Señora.

Portada del Álamo, in Fuentepelayo

In 1546 the second floor of the belfry arrived and the Chapterhouse Room and Library started to be constructed; in 1550 the bell tower was vaulted.

As a Gothic Cathedral geared towards the Renaissance and ending on the Baroque, it was thought about providing it with space to which end the City Council and the Council agreed to create a market in front of the main façade and to renew the prohibition of adjoining anything to the Cathedral on any side.

In 1552 the wooden and lead chapter which crowned the tower was finished and in the following year part of the old Cathedral was demolished once Carlos I had paid what he had promised.

In 1558 the remains and graves of Prince Pedro, Marisaltos and former prelates were transferred to the new Cathedral. In order to be able to use the Cathedral the transept was closed by means of a thick brick wall and on Toral Pillars in order to counteract the thrusts of the nave to this side until completing it.

The **second construction period** lasted from 1558 to 1577; in 1559 García de Cubillas died and so Rodrigo Gil de Hontañón was called upon once again and he continued the work until his death. On August 5th 1563 the first stone of the new chevet was laid. The City continued to donate money and wood from its Valsaín pine forests.

In 1570 what remained of the Romanesque Cathedral fell down and the subgrade was smoothed out to improve the surroundings before the marriage of Felipe II to Ana of Austria.

1562 saw the start of the construction of the Sacristy whose door is totally Renaissance and typical of Hontañón, being completed in 1572.

In the late 16th century the crisis set in, partly brought about by the conversion of the industrialists into stockbreeders who exported wool instead of working on it here, meaning that the works advanced more slowly; in 1575 Rodrigo Gil de Hontañón died, having lived to see the raising of the chapels of the ambulatory to the vaults' base and the guide base wall. **Ruiz de Chertudi** was appointed as the master builder though with duties as an architect.

In 1597 **Diego de Sisniega** took charge of the works. The crisis got worse and there was no recovery until 1605 when the economy was reactivated again. In 1607 **Pedro de Brizuela** took charge of the works, making the gateways of San Frutos and San Geroteo, and in the latter following the plan by García de Cubillas.

In **1614** a bolt of lightning destroyed the tower spire and a fire the tiles of the naves; Brizuela rebuilt it, respecting the Gothic buttresses and building what can be seen today within the most refined classicism of the age. He turned brick arches alongside the walls above the vaults in the central nave and the sides to increase the support surface of the roof beams, thereby relieving the Gothic wall and aiming his efforts directly at the pillars. In 1652 Francisco de Viadero lowered the projected height at the dome and replicated the crowning of the tower thereat.

This was the construction process for the last Gothic cathedral erected in Spain.

Great works of art are housed **inside** the cathedral, specifically for the new construction alongside works deriving from the former cathedral such as the **13th century Calvary** situated at the entrance to the Sacrarium chapel or the **choir stall**, carried out in the **15th century** commissioned by the bishop **Juan Arias Dávila**, to which seats of honour had to be added to adapt it to the new space. Worthy of special mention therein are the central seats taken by the Bishop and the monarchs Enrique IV and Juana of Portugal, each decorated with their respective coats-of-arms.

Another notable piece is the epistle **organ**, carried out in 1702 by **Pedro de Liborna Echevarría** which constitutes one of the most important pieces of this type in Spain.

The **retrochoir** is decorated by a retable with three bodies, whereof the central one was designed by Huberto Dumandre in the mid-18th century for the Riofrío palace[152] and donated by Carlos III; the side bodies were designed by Juan de Villanueva and carried out by **Ventura Rodríguez**. The central niche houses the silver trunk of San Frutos undertaken by Sebastián de Paredes in the 17th century; the finishing is raised thereupon with the sculptures of The Holy Trinity accompanied by St. Peter and St. Paul; on the side aisles the figures represent San Felipe and Santa Isabel in honour of the king's parents, Felipe V and Isabel de Farnesio.

The **Main Chapel** is totally devoid or ornamentation and its space was designed to be covered by a large Gothic retable which was never actually carried out as the works in this part of the Cathedral did not finish until the 17th century; in the second half of the 18th century the **retable** was commissioned to **Francisco Sabatini** who carried it out in marbles of various origins and colours at the workshops of the Royal Palace. This is a simple, yet refined and sombre retable, comprising three aisles in one body and an attic; in the centre there is the image of the **Virgin of Peace**, a 14th century work made from marble and silver which, according to tradition, was donated by Enrique IV. On the sides the stuccoed wooden sculptures of San Frutos and San Geroteo. In the attic, in the central part, the anagram of the Virgin, flanked by the figures of St.Valentine and St. Engracia.

Surrounding the Main Chapel is the ambulatory which has seven radial chapels instead of the five initially envisaged in the project. They are all decorated in line with the model of the chapel of San Frutos which was the first to be decorated.

As regards the chapels of the nave, it is worth mentioning those which house in their interior the most notable works of art.

The **Piety Chapel** was acquired by Juan Rodríguez, the churchwarden charged with supervising the works of the new cathedral. Here we can find the wonderful Renaissance work undertaken by **Juan de Juni** in the late 16th century; it is a retable in high relief on the theme of **Santo Entierro** featuring the supine body of Christ surrounded by

Santo Entierro de Cristo (Holy Funeral of Christ), Juan de Juni

the figures of the Virgin, St. John, Mary Magdalene, Mary Salomé, Nicodemus and Joseph of Arimathea. Juan de Juni carried out his artistic activity in various parts of Castile; firstly, he worked in León, Medina de Rioseco and Salamanca, later permanently settling in Valladolid where in around 1540 he formed one of the most important sculpture workshops of the Castilian school of the Renaissance. Despite his French origin and Italian training, he decided to settle here as did many foreign artists at the time owing to the attraction held by Spain at that time as an artistic centre. Part of his training took place in Italy and he travelled there as did many artists of the era; influences by Giacomo della Quercia and Michelangelo are notable in his work. By beholding his work we can assert that he is a totally mannerist artist imbued with highly defining characteristics such as the very pronounced pleats in the clothes, endowed with a lot of volume and in the majority of cases decorative as they take up too much importance in the composition; the bodies, the pleats and, generally speaking, the composition are defined by curved, winding lines which lend greater dramatism and expressiveness to the theme; the characters adopt very exaggerated gestures and very pronounced foreshortenings to emphasise the dramatism.

The **grille** which closes off the chapel derives from the Main Chapel of the old cathedral; it was carried out in the late 15th century by Francisco de Salamanca and is very similar to the grilles at Zamora Cathedral by the same author.

This then opens out onto the **St. Andrew's Chapel** where the **Triptych of the Descent** is kept, a work by **Ambrosius Benson**, a Flemish artist from the Bruges school and which derives from the St.Michael's Church[39]. In the centre of the retable there is a depiction of the scene of the Descent in which the characteristic of Flemish painting are more noticeable with the ample, pronounced pleats and the wealth of the fabric of the garments; on the sides there is a representation of St. Michael and St. Anthony, with the former following purely Renaissance canons. On the doors closing the retable there is a depiction of the scene of the Annunciation in grisaille.

The **Chapel of the Descent** houses one of the master works of **Gregorio Fernández**, his **Cristo Yacente** (Supine Christ), a 17th century Baroque work carried out in the last age of the Castilian master. The Supine Christ typology appeared in the late 15th century as a derivation from the theme of Santo Entierro, having been popularised by Gregorio Fernández during the course of the 17th century. This sculpture is characterised by its dramatism contained in the features, but highlighted by the blood flowing from the wounds; this is a slender body which has been treated with great naturalism, attention to detail and meticulousness, reinforced by the false eyes, teeth and nails which lend more realism to the image. It is endowed with those characteristics which define the work of Gregorio Fernández in the

individualisation of the locks of hair and the beard as well as the angular pleats of the loincloth and shroud. The image was intended to be viewed laterally on the seat of the retable of the chapel as a sculptural complement to the paintings of the central aisle which portray the Crucifixion and the Descent.

This is followed by the **chapel of Santa Bárbara** where we can find a **baptismal font** from the old cathedral, carried out in alabaster by the **Juan Guas** workshop whose vessel is adorned by the arms of Enrique IV.

The **Sacrarium chapel** takes up the place of the former Sacristy, to which the chapel has been added. The Sacristy was designed by **Rodrigo Gil de Hontañón** and it has a beautiful **retable** dedicated to the Christ of the Agonies, a 17th century work carried out by Manuel Pereira; it is framed within a retable made in ceramics by **Daniel Zuloaga** and its conception and colour recall the works by the Della Robbia brothers. The chapel, also known as that of Ayala Berganza, was built as a family pantheon by Juan de Ferreras at the Sacristy chevet; the beautiful plaster and brick **dome** is the work of **Juan de Setién Güemes** and it had to be reinforced in 1730 by a large buttress. The **retable** was designed by **José Benito de Churriguera** in the late 17th century; in the centre there is a tabernacle, the work of Antonio Tomé, carried out in the 18th century which comprises four faces, each having a relief alluding to the liturgical calendar which turns according to the time. It is crowned by a sculpture of Christ the Saviour.

The **stained-glass windows** entail an iconographic complement to the architecture of the temples, also lending its symbolic value of a religious nature. The vast majority of the stained-glass windows at Segovia cathedral were carried out in the **mid-16th century**, in line with the precepts of **Mannerism**; it was designed by Nicolás de Vergara, Nicolás de Holanda, Gualter de Ronch, **Pierres de Holanda** and **Pierres de Chiberri**, with the latter two being the most innovative and their works is characterized by large-scale compositions free of any architectural frameworks, experimenting with traditional techniques and using diluted colours and an abundance of grisaille and silvery yellow; its stained-glass windows constitute true paintings. The work by these masters was limited to the stained-glass windows of the naves whose main theme is the Redemption, with the arrangement of groups of three large windows, with the largest central one serving to portray scenes from the New Testament and the side ones depict foreshadowings from the Old Testament.

In the crosspiece there is a narration of the Life of the Virgin and in the ambulatory of the Public Life of Jesus; the stained-glass windows of the chevet were carried out in the 17th century by Francisco Herranz, recovering the technique. The stained-glass windows of the Main Chapel were removed by Sabatini when placing the retable to provide more light to this space, though they have been replaced during the course of the 20th century and they represent the Virgin, San Geroteo, San Remigio and San Alonso Rodríguez.

Cristo Muerto (Dead Christ) by Gregorio Fernández

In the late 15th century Bishop Juan Arias Dávila commissioned the construction of a new **cloister** to replace the old Romanesque one by **Juan Guas** who had already become a recognised master. This new cloister was **transferred** piece by piece to the location of the new cathedral between 1524 and 1528; the reason behind this move was mainly economic and the party responsible for the transfer was **Juan Campero** who made the base higher so that the cloister would be more stylised. The large windows of this cloister show tracery of rounded arches, harbouring other trilobed arches; combines S against S set, all enveloped by a pointed arch. The cloister is crowned by pinnacles and a fretwork parapet. Another work by the same author is the access door to the cloister; this follows the prototype model for the gateway by this artist to be found in other buildings of the capital such as the monasteries of El Parral[130] or Santa Cruz la Real[122] which is characterised by the overlaying of structures: basket arch with lintel on which there is a tympanum and all enveloped by a pointed, convex arch and with the last archivolt broken. The sculptures to be found on this door are in line with those which appear at the "Puerta de los Leones" (Lions' Gate) of Toledo Cathedral where he had been working. The sites for the sculptural decoration are usually always the jambs and the tympanum; here there is a depiction of 5th Agony or Piety which revolves more around the figure of the Virgin than around Christ. When the last archivolt is broken on the upper part, the arms of the Catholic Monarchs are usually gathered in the space formed. A characteristic aspects of the Guas Gateways is the contrast between the highly decorated area of the door and the smooth part of the wall which contains it.

The **cathedral museum** is located in the low body of the tower and in the Chapterhouse Room which are accessed from the cloister. In the centre of the first room there lies the **tomb of Prince Pedro**, the son of King Enrique II, who died - as tradition would have it – after falling from one of the fortress windows. Around the tomb there are many pieces of silverware and paintings, highlighting the triptych of the Virgin and Child, attributed to the Master of the Holy Blood, *La duda de Santo Tomás (the Doubt of St. Thomas)*, a work by Sánchez Coello or *El Cristo atado a la Columna (Christ bound to the Column)* by Morales.

One of the most emblematic pieces of the museum is the **processional monstrance** from the mid-17th century carried out in silver by Rafael González; it meets the prototype for an architectural monstrance crowned by a lantern on which you can see the triumphant figure of Christ the Saviour. Underneath there is a striking triumphal carriage from the 18th century made of gilt woodwork with a depiction of the Tetramorphs at the front part in the manner of conductors.

The **Chapterhouse Room** houses a wonderful collection of **Flemish tapestries** carried out in Brussels in the 17th century which almost totally cover the walls; here there is a narration of the history of Queen Zenobia. The composition of scenes is inspired by engravings from the time.

Alongside the Saint Geroteo gate there is an oil painting depicting the *Visit by San Francisco de Borja to Carlos V at the Yuste monastery*, undertaken by **Esquivel** in 1862 following the style called history painting which was so in vogue in the 19th century.

MARQUIS DEL ARCO HOUSE[21]

This was built in the **mid-16th century** in **Plateresque** style, becoming a model of this style in Segovia; it shares decorative similarities with the courtyard of the Torreón de Lozoya[30] palace, the other major Plateresque building in the city, though its design is more akin to the **Palace of Monterrey** de Salamanca and other Salamanca palaces in whose construction **Rodrigo Gil de Hontañón** was involved and to whom authorship is attributed of the palace we are looking at.

Since the 17th century it has belonged to the marquessate of Arco, but Felipe II had previously given it to his advisor, Cardinal Diego de Espinosa, who soon abandoned the property.

It has a simple façade which contrasts with the wealth of its interior **courtyard**; this is double-height with three porticoed sides following the Segovian typology. Both floors are architraed with base plates which support the structure which are richly decorated by friezes. However, what stands out at the courtyard are the **medallions** with very expressive busts of historic characters which fall within the line of works undertaken by Rodrigo Gil de Hontañón.

DIEGO DE RUEDA HOUSE[22]

It is also known as the **Álvaro de Luna house** though its presence thereat is not documented. The house dates back to antiquity as is borne out by the **Romanesque remains** which are still preserved such as the half point portal archway, now closed off, or the **tower** structure in the style of 14th century tower houses. However, the most remarkable thing about the building is its 15th century **courtyard** with ogee arch windows and open on three of their sides as is the case with Segovian courtyards. On the exterior the house opens out with an ogee archway crowned by the arms of the alderman Diego de Rueda and his wife Mencía Álvarez del Río.

CONVENT OF THE BAREFOOT CARMELITES[24]

Gateway of the Convent of St. Joseph

The Convent of the Barefoot Carmelites was founded by **Saint Teresa de Jesús** herself on March 19th 1574 at the Marquessate of Lozoya House; five years later they were transferred to their current location, near the previous one. The foundation mass was presided over by **Saint John of the Cross**[139] who was also the nuns' confessor during his three-year stay in Segovia.

This is a building endowed with a simple design whose façade boasts two gateways, one architraved whose origin must have been from one of the former palaces on which the Convent was built and the other, larger one provides access to the chapel, crowned by a triangular front broken to accommodate the relief of Saint Joseph with Child under a crown. Inside the chapel you will find a clearly Baroque appearance, highlighting the image of St. Joseph undertaken by **Luis Salvador Carmona**.

THE DEAN'S HOUSE[25]

This house is situated opposite St. Andrew's Church[54] and on the limits of the Canonjías[72] district whose name derives from its previous use; during the 17th century the house is said to have belonged to the bishop of Segovia Andrés Angulo. It stands out for its façade with a beautiful architraved gateway decorated by balls and framed by classical-style columns which leads us to think about a construction date of in around the early **16th century**. Above the gateway is the Bishop's coats-of-arms and a large **sundial**.

Access to the Calle del Corral del Mudo which still preserves its former medieval features.

ST. ANDREW'S[26]

This church is situated on the limit of the Claustra district and the former Jewish district[35] which had a synagogue on the adjoining square. The original appearance of the temple has changed a lot since its construction in the **12th century**; it was originally a Romanesque building with one nave and a semicircular apse, to which another polygonal was added on one side and a tower. In the early **17th century** it was submitted to a series of remodelling works which totally transformed its appearance, converting it into a temple with three naves when the porticoes closed and transforming them into the

Detail of tower frieze

Span of the central apse

lateral naves as well as covering the walls and ceilings with plasterwork. Hence, the original building can only be partly appreciated on the exterior, highlighting its brick **tower** and masonry, rhythmically decorated by the alternance of spans on its bodies and the contrast between light and shade on its brick friezes. The sculptural decoration of the apses has deteriorated greatly owing to the erosion of the stone and the motifs decorating the capitals of the windows cannot be appreciated clearly.

The interior of the temple, wholly Baroque, houses the main retable carried out in the first quarter of the 17th century on which **Gregorio Fernández** worked; he was responsible for the central high relief representing St. Andrew stripped of his clothes and the figure of The Saviour located in the attic. In both the master repeats the models and forms already used in his previous works, highlighting here the great dexterity with which he was able to overcome the problem of space in the relief.

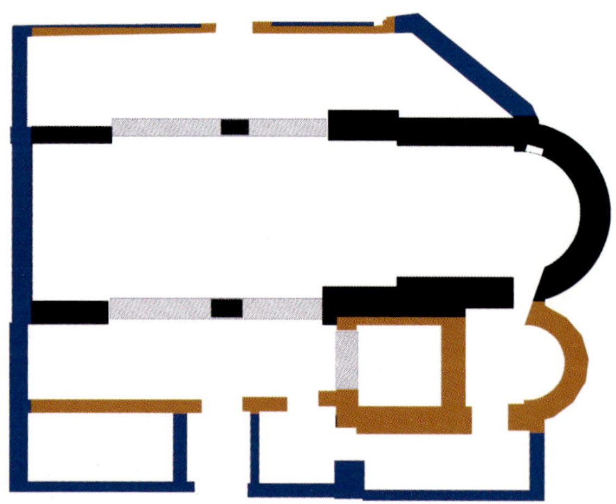

BUILDING CHRONOLOGY

- Romanesque 12th century
- Romanesque 13th century
- Baroque 17th century
- Removed during the Baroque modifications

THE SEGOVIA ALCAZAR 30

SHORT HISTORY

No-one knows much about the Alcázar until the 10th century, the time from when the first vestiges in the foundations date; it is named in 1122 in a document as a kastro (hill-fort) and in 1136 it is mentioned as an alcázar.

In 1158 Sancho III changed the location of the latter for some houses located in Almuzara, the high area of the City, with the Chapter of the Cathedral, though this change was later reversed; this must have been one of the first attempts at switching the fortress' location so that it was not isolated by the ecclesiastical district and improving its defences.

During the reign of Alfonso X, the Wise, the alcázar underwent several remodellings, some related with the fire apparently caused by a bolt of lightning in 1258, giving rise to the legend of Sala del Cordón (cord room). This same monarch created the Monarchs' Room and he commissioned the construction of the statues of the monarchs from King Pelayo to himself.

In 1322 withstood a siege, the upshot of the fight between the guardians of King Alfonso XI which spread chaos in the city of Segovia.

In around 1367 one of the best know legends about the fortress occurred, to do with the death of Prince Pedro, the son of Enrique II of Trastámara who, as tradition would have it, fell into the void from one of the windows in the Monarchs's room and his governess, fearful of the punishment she would receive for her carelessness, threw herself after him. With the Trastámara dynasty the alcázar was to experience, just like Segovia, its age of greatest splendour; it was no longer just a fortress, but it also became a regal residence and the custody of the royal treasure, the archive and the gunsmith's shop; this all led to the adaptation of the

Early middle ages foundations of a tower

spaces by means of remodelling work which granted it part of its current appearance.

Under Enrique IV the governors of the Alcázar came into his confidence, though without any political influence in matters of the realm. During the reign of this monarch and after his death, during the Civil War which fought against his daughter and his sister, the alcázar became a strategic point to be taken control of by the bands confronting each other and its governor, Andrés Cabrera, played a very important role within the interplay of interests.

With the arrival of the house of Habsburg to the Crown, the alcázar became to gradually lose its importance and became just a state prison.

During the reign of Carlos III, within its reformist programme, the Artillery Academy was founded here where it remained for a century until it moved to its current location in the convent of San Francisco after the fire suffered by the Alcázar. A lot of damage was caused by the fire and a long restructuring process was required until attaining the image it has today; it currently houses the General Military Archive and it is administered by the Alcázar Trust, founded for this purpose.

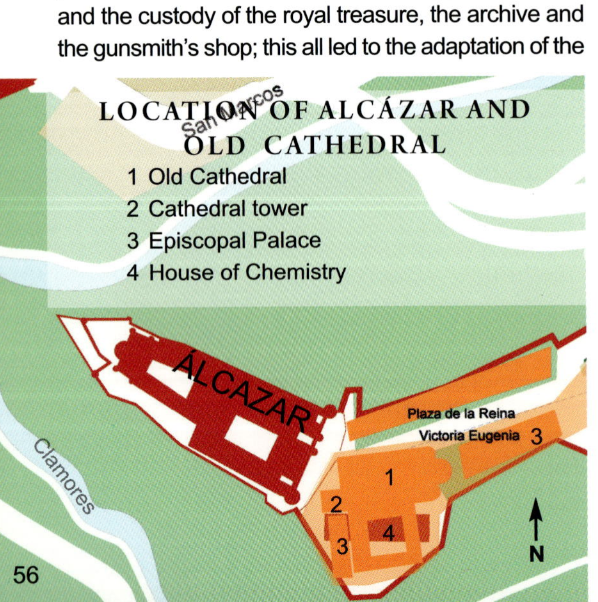

The alcázar from St.Mark's

HOUSE OF CHEMISTRY

Alongside the Alcázar, on a part of the plot of the former cathedral, you will find this Neoclassical building designed by Sabatini; it was named after him as he was the first director of the laboratory of the Artillery Academy Joseph-Louis Proust who discovered the law of definite proportions here whilst carrying out his teaching activity.

MONUMENT TO DAOÍZ AND VELARDE. This monument is located at the centre of the gardens of the Reina Victoria Eugenia square, commemorating the **Heroes of May 2nd 1808** at the Monteleón Barracks against the French invaders who, what's more, were students of the Royal Artillery College. The monument began to be built in 1908, coinciding with the 1st Centenary of the feat and it was inaugurated in 1910 by King Alfonso XIII; its author, **Aniceto Marinas**, made of use of various materials to carry it out: bronze, granite and marble.

The monument has three heights, on the lower one there is an allegory of the story characterised as a woman with a book who is noting down what happens in the interim period which captures the dramatism of the struggle; crowning the whole, there is a woman representing Spain with a flag, receiving the bodies of the two heroes and under them the inscriptions reads:

TO THE ARTILLERY CAPTAINS MR. LUIS DAOÍZ AND MR.PEDRO VELARDE THE SPANISH NATION.

BUILDING CHRONOLOGY

The oldest parts of the Alcázar of Segovia seem to date back to Roman times which is the case of the ashlar of the square tower of Liza (battle) though its dating has not been reliably confirmed, in the worst case scenario it would have an indefinite Early Middle Ages dating.

Of that part preserved, the oldest part is the heart of the castle, the space of the parade ground and the enclosure walls as far as the forecourt as well as, perhaps, a fence which would reach as far as the tower of homage.

There has been speculation about the notes of a medieval visitor regarding the possible existence of another put which would separate the parade ground area from the clock, though it has not been possible to confirm this even with archaeological excavations.

In all cases the original castle from the 10th century would be made up of two square towers and another main one where the access to the courtyard would be housed in an angle in Islamic fashion. Later, perhaps simple walls would contain temporary constructions for accommodating troops and supplies. In the 11th century the Wall had already been built which closes off the Old Palace room where there still remain two spans and several mullioned windows, all Romanesque.

Juan II tower window

In the 13th century, still during the Romanesque period, a major extension to the castle was carried out; to the north, vaulted basements were built, capable of housing the rooms of the throne, the Galera and las Piñas, and so the mullioned spans which looked out onto them were removed and new ones were opened up onto the Eresma valley. To the south, the wing was raised which is today taken up by the Artillery Museum and it was joined to the Old Palace by means of the wing which adjoins the Juan II tower and on whose walls there are Romanesque spans.

Furthermore, the defences were strengthened with the increasing of the Juan II tower which forms an unicum from two preceding ones and whose point of joining is still visible on the lower part.

Two new towers were built at the ends of the pit which are joined to the two previous ones which had

BUILDING CHRONOLOGY

1 Aqueduct
2 Early Middle Ages or before
3 Juan II tower
4 Weapons Courtyard
5 Horses Room
6 Chimney Room
7 Throne Room
8 Galera Room
9 Piñas Room
10 Royal Chamber
11 Monarch Room
12 Cord Room
13 The Queen's Boudoir
14 Chapel
15 Sundial Tower
16 Weapons Room, Tower of Homage
17 Treasure Room
18 Sundial Courtyard
19 Artillery Museum
20 Shop

Romanesque 11th century
Romanesque 12th century
First half of the 13th century
Second half of the 13th century
15th century
Gómez de Mora, 16th century
Other periods

FOSO

already advanced the defences as far as the very edge, creating a larger Liza. The reasoning behind this reinforcement becomes obvious if it is considered that the old Cathedral of Segovia was raised opposite Juan II tower and its bell tower was in competition with it.

To the west the walls joining the Tower of Homage to the rest of the palace were raised or strengthened and the chapel was erected. The well courtyard and the nearby watchtower of Polvorín took on their current configuration.

In the second half of the 13th century the Royal Palace was extended again with the addition of the Monarchs' Room, erected again on the vaulted basements and with a north wall which must take its thickness from the decreasing configuration of the site.

Finally, in the 15th century, the rooms called El Cordón (The Cord) and El Tocador de la Reina (The Queen's Boudoir) were erected.

As regards the Juan II tower which had been increased, it again gained further height with the arrival of the Trastámara dynasty; the window discovered recently dates back to this time and it has a strong Islamic influence. Enrique IV, judging by the pomegranates which decorate it, dressed it up and provided it with watchtowers features and its design is attributed to Juan Guas, having

Succession of constructions
Rock bed,
Gothic fence
Base of the Romanesque tower
Coronation of the Catholic Monarch
Flemish roofs
Aqueduct, under the drawbridge.

been finished during the reign of the Catholic Monarchs. At the same time the Catholic Queen had the chapel vaulted with Gothic vaults, they were lost during the fire in 1862.

In 1521, during the Guerra de las Comunidades (Revolt of the Comuneros), the Alcázar suffered serious damages as it was laid siege to by the Comuneros who lodged in the Cathedral; as a consequence, the Cathedral was transferred to its present plot and in the Alcázar, as well as repairing the damages, the lower quarters of the Juan II were packed solid to strengthen it.

During the reign of the apathetic Carlos I nothing was done to the castle, whilst with this son Felipe II, the Architect King, the building attained its present condition; commissioning the remodelling works to Gómez de Mora, the brilliant classical courtyard was built, replacing the medieval one and which, increasing the length, respecting the local custom of leaving the north side without any gallery where an elegant curtain wall was placed to cover up the medieval one. In the same way, the courtyard steps were carried out, that at the Patio del Reloj (Clock Forecourt) and those at the Tower of Homage which were finished at a tower with classical buttresses until the fire in 1862. Opposite the alcázar, towards the forecourt, the so-called Galería de los Moros (Moors' Gallery) which was a covered bailey for the artillery and which dismantled after the fire. It was put back together and the iconographic programme of the Monarch Room was finished, adding the figures of the queens and finishing the iconographic series with Queen Juana I. And, first and foremost, everything was bestowed upon the slated roofs like those

that the King had seen in Flanders during a trip and which he had liked so much; with this in mind, he commissioned specialized slaters and had them bring the slate, on his own instructions, from the outskirts of Santa María la Real de Nieva; the work started in 1559 and it was completed in 1562. The works and design were taken on by Gaspar de Vega who also urbanised the forecourt bordering the Alcázar.

At this time the work to deepen the pit must have been completed, i.e. the pit which received the drainage channel on an indefinite day in the 17th century.

On June 10th 1681 a bolt of lightning struck the Tower of Homage which destroyed the roofs in said area, though they were quickly rebuilt.

And finally on March 6th 1862 a greenhouse located in the Queen's boudoir produced the wood of the ceilings and the fire spread throughout the Alcázar. The fire consumed everything, floors, coffered ceilings, works of art,...

After several years of ruin and negligence it was decided to start the reconstruction to which end the excellent and detailed drawings by José María Avrial which allowed the recovery of the coffered ceiling drawings. The original friezes were restored and they still remain and all vestiges of Romanesque were discovered which are spread all around Alcázar. The roofs were raised once again as they had before and some Neogothic details were added such as the windows of the Royal Palace and of the chapel.

In this way, it was managed to recover the splendour of that which was lost barely without distorting what it had been before, providing old elements which had been concealed and reinforcing the reading of its history.

ARMAS COURTYARD

The courtyard designed by Francisco de Mora combines the traditional northern roof in order to keep it from the cold with the extreme elegance and simplicity of the Works of Spanish classicism. Square selection pillars support slightly moulded arcades on which a wainscot runs which supports the upper flooring, again arranged in square pillars though because it has not upper floor, finished in monolithic lintels. The parapets are bedecked in quadrangular boards. On the fourth side the whole layout is repeated and the diagram for the other three sides, but in this case closed. The parade ground is crowned by walls with pilasters and thermal spans.

The **Sundial Courtyard** goes by this name because of the sundial situated on one of its walls. The coat-of-arms featuring on the façade belongs to Emperor Carlos V and it was brought from St. Martin's gate when the latter fell down.

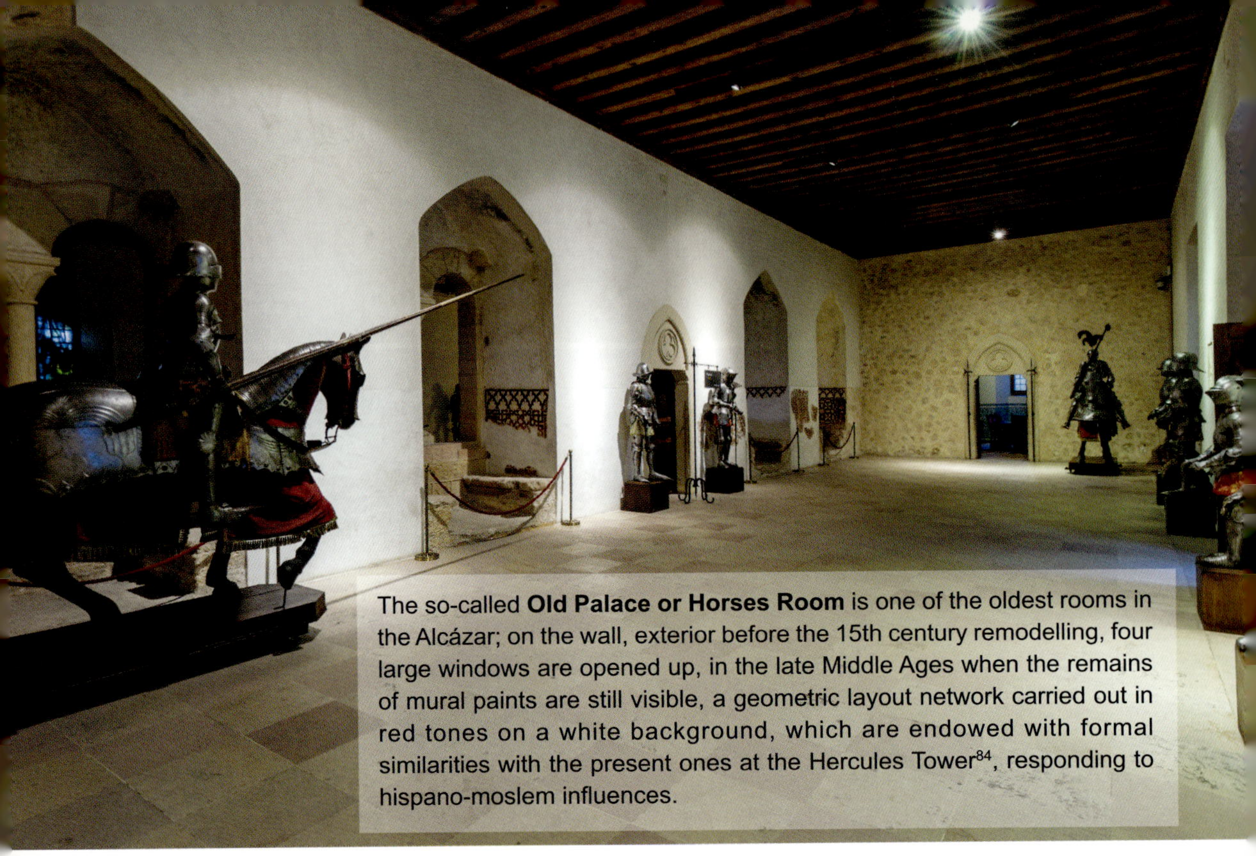

The so-called **Old Palace or Horses Room** is one of the oldest rooms in the Alcázar; on the wall, exterior before the 15th century remodelling, four large windows are opened up, in the late Middle Ages when the remains of mural paints are still visible, a geometric layout network carried out in red tones on a white background, which are endowed with formal similarities with the present ones at the Hercules Tower[84], responding to hispano-moslem influences.

The **Chimney Room** forms part of the so-called Old palace and it provides access to the Throne Room of the extension; it has a square made in Talavera ceramic and their walls are decorated by various paintings, highlighting the portraits of Felipe III and Isabel Clara Eugenia.

The **Throne Room** affords communication with the former medieval palace; its Works were carried out in the mid- 15th century as part of the project to extend the palace to the north. The construction of this room was commissioned by Enrique IV as stated in the inscription which runs along the plaster frieze of the high part of the wall; this frieze is original, but it was highly damaged; even so, part of its decoration based on strips can be made out with vegetal and animal decoration which are intertwined, forming circles. Another wooden frieze runs on it with mocarabe decoration which, just like the original housing, was destroyed by fire; the current reinforcement derives from the church of Urones de Castroponce in the province of Valladolid as this coincided with the original in terms of size and decoration.

Throne Room

Galera Room

The **Galera Room** is part of the extension and it was built by the widow queen Catherine of Lancaster. On its southern wall the Romanesque windows of the room of Palacio Viejo open out and you can also see the sgraffito which adorned the old façade. This room was greatly affected by the fire and the ceiling is a faithful reconstruction from the drawings carried out by José María Avrial. Under the coffered ceiling there is a plaster frieze endowed with three strips, with the central one occupied by fine geometric decoration of hispano-moslem roots and coats-of-arms from Castile and León; the end strips bear an inscription, the upper one referring to the works promoter. Above the frieze, the Mocárabe decoration serves as a conduit to the coffered ceiling, decorated by eight-point stars with vegetal decoration and pineapples in the centre.
The mural adorning the back wall represents the proclamation by Isabel I at St. Michael's Church[39] and it is the work of Carlos Muñoz de Pablos, also the author of the stained-glass windows portraying Enrique II and Enrique III along with his wife Catherine, the promoter.

The **Piñas Room** takes its name from the decorative element which is repeated alternately on the wooden ceiling.
The plaster frieze, running along the high part of the wall, is the original commissioned by Enrique IV, however, the ceiling is a reconstruction of the original from the drawings by Avrial. The frieze has a fine vegetal decoration framed in quadrants in which Islamic-rooted elements alternate with others of a Gothic nature; on some caryatid angels with the coats-of-arms of Castile and León.
The **Royal Chamber** belongs to the old palace; its doors are a copy of the those which used to be at the Royal Palace of St.Martin[26], commissioned by Enrique, and they derive from an individual who donated them to the Alcázar.

Piñas Room

The **Monarch Room** began to be decorated in the time of Alfonso X and he commissioned the depiction of the **monarchs of Castile** from King Pelayo to his father Fernando III; in the 15th century Enrique IV continued the series all the way up to himself, having been definitively finished by

Felipe II, who undertook remodelling work and included the queens, the Catholic Monarchs and Juana I, as well as the four counts: Fernán González, the resettler part of the province of Segovia around Sepúlveda, El Cid Campeador (the Champion), Raymond of Burgundy, the resettler of Segovia and Enrique de Lorena.

This room was the place chosen to celebrate the civil marriage ceremony between Felipe II and his niece Ana of Austria.

It was also greatly damaged in the fire, having remade the royal figures from the drawings by Avrial.

The **Sala del Cordón** (Cord Room) was commissioned by Enrique IV, though tradition relates it with Alfonso X who had it decorated with the Franciscan cord which is laced around it after a fire caused by a bolt of lightning that devastated part of the alcázar and attributed to a divine punishment for the king's arrogance. The next room is the one which is believed to have been the **Queen's Boudoir** where the 19th century fire started that put paid to the ceiling decoration and the furniture.

The existence of the **Chapel** seems to date back to an oratory of Romanesque origin which was remodelled by Isabel the Catholic in the late 15th century. The armour comes from the Segovian village of Cedillo de la Torre and it is a 15th century work, whilst the retable comes from Viana de Cega in the province of Valladolid. The painting representing *The Adoration of the Magi* was carried out by **Bartolomé Carducho** and it was one of the few goods saved in the fire thanks to the intervention of the cadets from the Academy who cut it out of its frame and took it away.

The **Weapons Room** and the **Treasure Room**, the grille closing the door of the Treasure Room is of Romanesque origin; in the centre of the room there is a press from the Royal Mint.

A collection of arms is displayed in which the crossbow of Carlos V is worthy of special mention.

Gunsmith's

THE CLOISTER OR CANONJÍAS[31]

Romanesque gateway c/ Daoiz, 16

Romanesque gateway c/ Daoiz, 18

Romanesque gateway c/ Daoiz, 21

This is an urban fabric endowed with great historic and artistic value, bearing testimony to a way of life. Claustra (Cloister) or Canonjía (Canonry) is the name given to the **district** formed in the environs of the **old Cathedral** where the dwellings of the **canons** were situated; nowadays it is one of the few examples of this type of district which are preserved. It is situated in a well-defined area on some sites that the Council donated in around 1120 to the bishop and to the cathedral which are limited by the city walls and they stretch from the St. Andrew's Church[54] to the Alcázar[56]; the enclosure remained closed during the nights by **three doors**, only on which remains on Calle de Velarde, with the other two being situated at Calle de Daoiz and at Plaza de Juan Guas. This space constituted a **small city** within it and it had its own rules and regulations, with the canons being governed by conventual rules. The situation changed after the Revolt of the Comuneros with the demolishing of the old cathedral and the construction of the new one, many parish priests starting living outside its walls and the district lost its purpose.

All the dwellings in the district are structured in the same way; on the street a two-height façade with

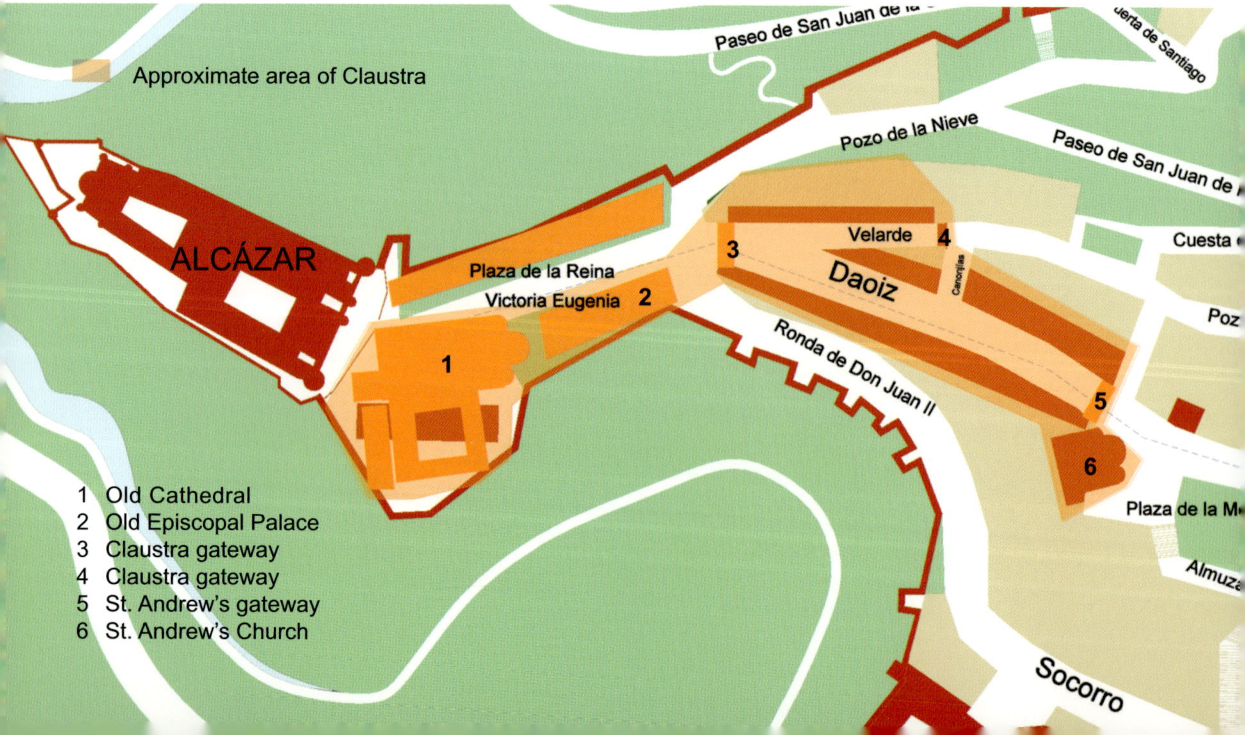

Approximate area of Claustra

Paseo de San Juan de la
Puerta de Santiago
Pozo de la Nieve
Paseo de San Juan de
ALCÁZAR
Plaza de la Reina
Victoria Eugenia
Velarde
Daoiz
Canonjías
Cuesta
Ronda de Don Juan II
Plaza de la M
Almuza
Socorro

1 Old Cathedral
2 Old Episcopal Palace
3 Claustra gateway
4 Claustra gateway
5 St. Andrew's gateway
6 St. Andrew's Church

Claustra gateway at Velarde street

few spans and a typical Romanesque gateway that connected with the entranceway. The lower storey was taken up by those spaces dedicated to the service whilst the upper storey constituted the main floor; all of this centred around a courtyard. At the rear the garden opened out with a garden area and here, owing to difference in level with regard to the street, there was a gallery and a winery. All the houses had water from the aqueduct channels which was stored in cisterns.

These houses constitute one of the best examples of civil Romanesque architecture owing to the good state of repair of the volumes and the decoration, largely due to the fact that it has belonged to the council until recently. Those which retain more original elements are those located at Calle de Las Descalzas nº1, Calle de Daoiz nº 18 and, in particular, nº 19 thereof, known as the Casa de Argila.

Romanesque gateway c/ Velarde, 20

ANTONIO MACHADO HOUSE~MUSEUM[32]

Antonio Machado arrived in Segovia in 1919, residing during his stay in the city in the house currently taken up by the museum and which, at that time, was a boarding house run by Mrs. Luisa Torrego. He was there until 1932, holding the Chair of French at the General and Technical Institute; during this stay in Segovia, as well as undertaking prolific literary work, he also got involved in the cultural activity of the city, attending literary circles, giving conferences, writing articles and being one of the founders of the "**Universidad Popular**". It was here in Segovia where he wrote the beautiful poems dedicated to his beloved **Guiomar** who he got to know during his visits to Madrid. This one-time dwelling of Machado in Segovia was acquired by the **Royal Academy of History and Art of San Quirce**, the successor of Universidad Popular, which has compiled photos, books, articles, paintings and personal objects of the famed writer which are today on display at the museum, recreating his life; worthy of special note is the bedroom with its bed, table and heater.

CASA DE LOS LINAJES [33]

This old **Romanesque house** was built on the way to the Romanesque church of San Pedro de los Picos and today the only remnant of this past is its **doorway** as the building was remodelled in the 15th century, at which time it had the alfiz which frames the doorway, with the arch tympanum and coat-of-arms of its owners, the Falconi family, coming later in the 16th century.

SAN PEDRO DE LOS PICOS [34]

This **mid-12th century** church, whose original dedication was to San Pedro ad Vincula, takes its more popular name from the beak-shaped finishes which used to crown its tower. The temple has a nave finished off in a **semicircular apse** adorned with cantilevers on the exterior and on the northern side its retains its round arch doorway adorned with flowers contained in circles. For a long time it remained abandoned and in a state of ruin until in 1961 the Count of Melgar acquired it under a 50-year concession, undertaking works to restore it and adapt it to a painting studio.

HOSPITAL DE LA MISERICORDIA [36]

The hospital was founded in the late **15th century** by **Bishop Juan Arias Dávila** to attend to the poor sick, but its construction was not started until a century later. Apparently, the Church design was provided by **Rodrigo Gil de Hontañón**, whose creative genius can be seen in the church doorway which follows the composition lines characteristic of the master, highlighting the Baroque image of the Virgin of Mercy. Alongside this hospital are the ruins of the former Convalescents' hospital, raised in the early 17th century to shelter the sick which come from the previous one, but which had not yet been fully recovered. The chapel design was given by **Pedro de Brizuela** who was also working on the previous church.

ST.STEPHEN[39]

This Romanesque church forms a fundamental part of Segovia's profile thanks to its very high **tower** which actually competes with the Cathedral[40] and Alcázar[56] themselves. However, the original appearance of the temple has changed greatly over time, including its tower which in the early 20th century underwent intense restoration of a historicist nature. The putter appearance of the Romanesque church was changed radically as from the 15th century when chapels attached to its northern wall began to be added to accommodate the burials of numerous noble families of the city. However, what undoubtedly altered its appearance most was the **Chapel of Peace** whose construction was ordered transversally to the transept in the 18th century. All these constructions had a major effect on the internal layout, increasing the Romanesque walls and covering the ceilings with Baroque plasterwork. The former northern apse houses the beautifully crafted image of **Christ on the cross** deriving from the extinct church of Santiago and carried out in the 13th century.

As we have indicated, the most noteworthy element is the **tower** which almost disappeared owing to its poor state of upkeep, exacerbated in 1894 when a bolt of lightning fell on its now gone Baroque slate spire. The restoration stretched out over time, ending in 1922; during the course of the works the portico was also affected as the tower scaffolding fell down after a strong gale which totally destroyed it and it had to be rebuilt piece by piece. The restoration of the tower was carried out in line with the criteria of the time, very different from the current ones, in such a way that the vast majority of the spires were replaced with new ones, distributing the original ones around different monuments and buildings in the city as we can see in the Church of San Quirce[80] or at a house on Calle Daoiz[72], with their remains to be found at Segovia Museum[6]. The structure of the tower denotes rhythmic treatment of the architectonic elements; it has a blind base up to the height of the nave and from then onwards

five bodies, with the lower two being endowed with two blind spans on each face and a greater height for the following two bodies, with the latter opened with dual spans which are smaller than the previous ones; the top floor opens out with three spans and is less high than the previous ones. This solution creates an optical effect of greater height than the real one.

The **porticoed gallery** is supported on dual columns whose spires denote a very deteriorated appearance and some scenes can still be made out relating them with the porticoes of San Lorenzo[118], San Martín[32] and San Juan de los Caballeros[88], all influenced by the Duratón[155] church gallery.

EPISCOPAL PALACE DIOCESAN MUSEUM[40]

The building was acquired in the mid-18th century by the Bishop of Segovia, Manuel Murillo y Argaiz, who was granted it by the Salcedo family, the former owner of the property.

The building started to be constructed in the **mid-16th century** and its monumental façade dates back to this time, the best Renaissance example in Segovia; it is endowed with a beautiful gateway framed by fluted columns which support a triangular front whose tympanum encloses the coat-of-arms of the Bishop propped up by two bare male figures. At the spandrels of the access arch and in the keystone thereof three figures appear in relief who represent **Hercules**, the mythical founder of Segovia at different times of his life; hence, in the centre he appears with the snake sent by Hera when he was a boy, on the left spandrel carrying the two columns which separate Europe and Africa and on the right breaking the jaws of the lion of Nemea. This classical-style repertoire is complemented on the rest of the façade by garlands and triangular fronts with Roman busts in the windows of the lower floor.

After the acquisition of the building by the Bishop the works were undertaken to finish it and the sober double-height courtyard dates back to this time, merely adorned on the cornice by fronts shared with the coat-of-arms of the Bishop.

It currently houses the **Diocesan Museum** which is complemented by the **ceramics' collections** of the **Zuloaga** family and the **glass and crystal collection of San Ildefonso** from the Laguna Lomillos couple which brings together pieces from the creation of the Royal Factory[150] until the 19th century, with both collections comprising more than 300 pieces each.

SECRETARY'S HOUSE 41

Known by this name as this was the residence of Gonzalo Pérez, the secretary of Felipe II; the advisor to King Enrique IV lived here previously. The most noteworthy element of this building is its **gateway**, a **16th century** work attributed to **Rodrigo Gil de Hontañón** who was working at the Cathedral at that time; it is a half point portal archway framed by fluted semi-columns on which a frieze of grotesques runs which is repeated under the balconies. The busts are situated at the spandrels, one female and the other male, identified with Hercules; a triangular front opens out onto the frieze between candelabras accommodating the family coat-of-arms sustained by putti. The decorative resources are frequently used in the 16th century and they attain their maximum expression in Salamanca by Rodrigo Gil de Hontañón.

MÍNIMOS CONVENT MIÑÓN ART SPACE 42

The **former church** of Our Lady of Victory of the former Convent of Franciscan Minimos was erected in the late **16th century** following the Baroque models of the time, though, at present, it looks quite different; after the monks had left it was used as a **theatre**, known as the Main Theatre, of Victory or Miñón which resulted in the transformation of the internal appearance of the chapel, with the nave of the church becoming a courtyard of seats, the transept a

stage and adding high boxes. It has remained with this appearance until today, fusioning different elements of its two former uses, giving rise to the only space that has been restored, retaining its personality and its historic course to accommodate a restaurant and a contemporary art centre.

ST.QUIRCE[43]

During the course of its history this temple has undergone numerous hardships as in the mid 19th century it ceased to serve as a place of worship and it became a warehouse and barn until in 1927 it was bought by **Universidad Popular Segoviana**, founded in 1919 with a view spreading learning and its members included Antonio Machado[74]; in 1955 the Universidad Popular became the **Academy of History and Art of San Quirce**, the current church owner, thanks to whose work it has been managed to preserve and improve its appearance. As far as the church is concerned, this was built in the late **12th century** and early 13th century in two successive construction campaigns, with the essential part of the temple belonging to the former with its nave and semicircular transept where we can find an extremely interesting **window** which bears decorative elements relating it with other temples in the city such as St. Millan[114], San Sebastián[95], San Juan de los Caballeros[88] or that near La Trinidad[82], all influenced by the workshops of Ávila; it should be pointed out that a consequence of its misuse as a barn, this window was seriously damaged and it was thus rebuilt using the surplus material from the restoration of St. Stephen's[76] tower. During the second construction era the tower was erected with its chapel, presenting the form of a semicircular apse to the exterior.

CONVENTO DE LAS OBLATAS[38]

Located opposite the transept of the church of San Quirce you will find this **17th** convent which initially belonged to the Order of Hooded Franscisans and which, after many hardships, was passed on to the Religious Onlations' Order who abandoned it a few years ago. Their outer appearance is very simple, only the church doorway adorns with an image of **San Buenaventura** flanked by the coats-of-arms of the Maecenas works' patron.

ST. NICHOLAS[46]

In the close vicinity of the churches of St. Cyricus[80] and La Trinidad[82] you will find this **12th century Romanesque** church is located, having lost its parochial dignity in the late 19th century; it currently houses the head offices of the Municipal Jesus Theatre, owned by the Council which also uses it as a cultural centre. During the course of its history it has undergone numerous interventions and its demolition was even contemplated though it has managed to be maintained until our days very transformed. Of its Romanesque origin the transept has been maintained and at a later period, but also at the same time, are the tower and its **funeral chapel**, similar to those of San Clemente[112] and St. Cycicus[80]; the nave was redone as was the portico which during the restoration in the mid-20th century received the columns of the palace courtyard from Arias Dávila[86]. Inside, in the side chapel, the remains of **mural paintings** are retained around an arcosolium whose iconography is related with the funerary nature of the room and in the interior of the tomb there is a depiction of the funeral with members of the clergy and on the exterior the ascent of the deceased's soul carried by angels; although the pictorial technique used is different, both scenes can be dated back to the 13th century within the style called **Linear Gothic**.

SANTÍSIMA TRINIDAD[45]

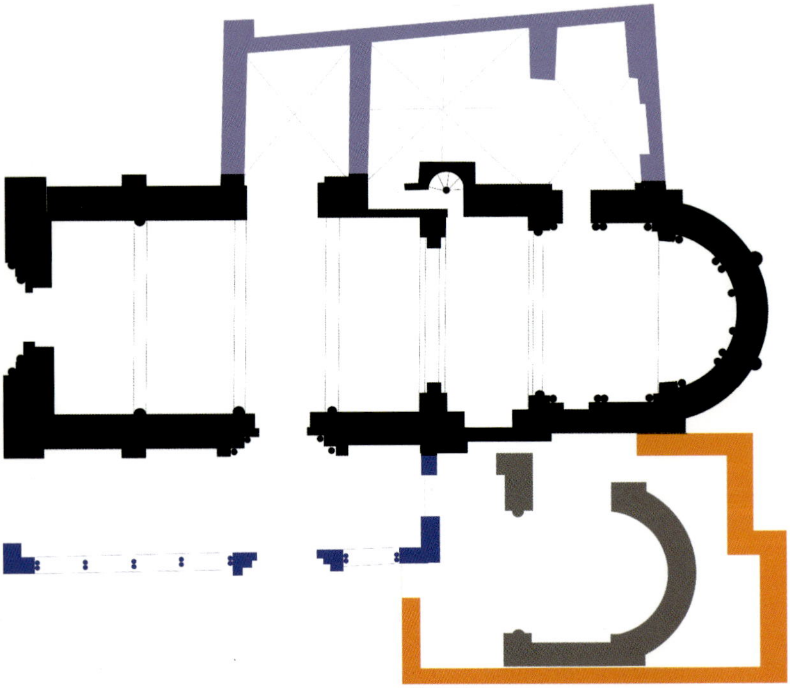

CHRONOLOGY OF THE CONSTRUCTION

- Previous church, 11th century, remains
- Current church, 12th century
- Late Romanesque portico, 13th century
- Campo Chapel, 15th century
- Modern

This church, founded according to tradition by Díaz Sanz, one of the leaders of the reconquest of Madrid from the Moslems, was erected in the **mid-12th century** along with the remains of a **previous temple**, also Romanesque, dating back to the late 11th century or the early 12th century, smaller in size, located on the southside of the current one and whose vestiges were used as a chapel.

This is a church with only one nave crowned by a **semicircular apse** whose exterior bears beautiful spires on its windows with vegetal and fantasy scenes; in the interior, the apse's configuration reminds us of the church of St. Milan[114], articulated by a blind arcade whose spires have been greatly restored owing to the intervention undertaken in the mid-20th century. Also noteworthy is the chapel of the Campo family whose house was constructed opposite the portico with a late Gothic gateway.

On the **exterior** there are two similarly constructed gateways, with the southern one showing on one of its spires – slightly roughly finished - the scenes of the Visitation and Nativity, the latter with a curious treatment of perspective. The porticoed gallery protecting it was raised on the nave of the previous church and it show similarities to that of the church of San Clemente[112].

TOWER OF HERCULES, SANTO DOMINGO EL REAL CONVENT[44]

This is the nickname given to the **mid-13th century Romanesque** palace which currently occupies the Convent of Dominican Mother Superiors; it was originally called the **Palace of Don Alimán**, but as from the 16th century it began to be called the Tower of Hercules as this historical figure was regarded as the founder of Segovia and owing to the presence of a granite **structure** which has been interpreted as the third labour of Hercules: the capture of the Erymanthian boar alive, an animal which wreaked terror in its region and which was caught by the classical hero. In the sculpture, which is semi-built into the wall, Hercules can be seen resting his foot on the head of the beast as a sign of submission. Access was gained to the palace via a gateway, now closed off, opposite the church of La Trinidad[82] and alongside the current 17th century convent chapel, the work of Pedro de Brizuela. The most characteristic element of the whole is undoubtedly its round-looking medieval tower whose interior conceals a series of very interesting medieval **murals**, located on the baseboards of the first two floors; they are painted in red on white tones and they represent a multitude of intertwining geometric shapes of Mudéjar tradition, similar to those of the Alcázar[56] and figurative battle scenes, as well as enigmatic inscription in kufic characters.

Around the church of La Trinidad[82] we can find numerous vestiges of **Romanesque and Gothic civil architecture**, reminiscent of the important medieval city that Segovia used to be. Opposite the southern gateway is the **Campo House**, rebuilt in the 16th century, in which the loopholes of the façade can still be seen, bearing testimony to its medieval past. In Guevara square there are two buildings featuring remains from that time, on one side, the **Ramírez de Arellano palace** where, alongside the current gateway after the deteriorated plaster, part of its former Romanesque entrance can be made out; and, on the other side, on the narrow street that continues as far as the following square, the remains appear of a brick gateway with a semicircular archway.

Tower of Hercules from the Convent garden
Photo: courtesy of the Community of Mother Superiors of the Convent of Santo Domingo el Real de Segovia.

PALACE OF THE COUNTS OF MANSILLA[47]

This is undoubtedly one of the most important masters of civil medieval architecture; as with the other buildings from that time, it has undergone numerous remodelling work to adapt it to its new uses and so alongside the Romanesque construction there are Gothic and Baroque additions. As far as the Romanesque remains are concerned, found during some restoration work, worthy of special mention is the 13th century **arcade** and a **vaulted room** with brick as well as gateways which are also made of brick. This space was filled and used as the foundation for the 15th century **Gothic courtyard** built here and which is structured into floors, the first endowed with granite columns and spires emblazoned with the arms of prominent lineages; the upper floor is wooden and is supported on footings made of the same material.

In the 17th and 18th centuries a series of remodelling works were carried out to extend the property, giving rise to the current façade crowned by a back-comb with the heraldic shield and opened by a lowered archway.

Romanesque gateway

Romanesque arcade

Arias Dávila tower

VILLAFAÑE PALACE$_{48}$

This **mid-16th century** building boasts a totally Renaissance façade with a gateway framed by columns finished off by candlesticks; on it, inside an oval bas-relief and amongst vegetal decoration, there lies the shield of the former owners of the property.

Worthy of special mention in the interior is its **courtyard** and some **murals** in grisaille from the same time as that when the house was built.

ARIAS DÁVILA TOWER$_{49}$

This was raised on a previous Romanesque construction of which there are still some remains; its promoter was **Diego Arias Dávila**, the accountant and scrivener of Enrique IV who had it built in the **mid-15th century**. This type of construction was typical of the period as noble families usually had palaces in the cities and would convert them into true fortresses to demonstrate their power; many of these towers disappeared during the reign of the Catholic Monarchs who cut the tops off them to punish those nobles who had not supported their cause, symbolically putting a brake on their power. There are still many good quality examples in Segovia of this type of construction, highlighting, along with the one we are looking at, the Lozoya tower[30] and the tower of Hercules[84].

The **palace** accompanying it is a **16th century** work which was built by **Pedro Arias Dávila**, the son of the former. The unit as a whole has been changed greatly over the centuries and the tower is the best preserved element with its defensive merlons as the rest of the palace has been mutilated to adapt it to new uses with the disappearance of another tower located on the main façade as well as the battlements which crowned this side; the current gateway derives from another palace and the columns of the courtyard were transferred to the portico of the church of St. Nicholas[81] where they can be seen today.

Villafañe Palace Gateway

NOBLE HOUSE~RODERA ROBLES MUSEUM 50

This building from the late **15th century** belonged to the lineage of the Contreras as denoted by the small shield located on its gateway. In the mid-20th century it housed the Provincial Museum until the latter was transferred to its current site at Casa del Sol. At present, as well as recreating a typical dwelling from the time of its construction, it accommodates the Rodera Robles Museum with a vast collection of ceramics, glass pieces and paintings where we can find works by major artists such as **Beruete, Fortuny, Eduardo Vicente** or **Sorolla**; all these objects belong to the personal collection of the couple formed by Mr.Ángel Rodera and Mrs. Rafaela Robles. The museum also has a space reserved for **Engravings** where the working tools and the different techniques are shown, taking a tour around their history.

This church is situated alongside the city wall and it is regarded as **the oldest one** in Segovia and one of the most important ones as it was the seat and place of burial of the **Noble Lineage Board**. There has been much speculation about its origin and some people had suggested an original Visigoth construction, later Mozarabe, whose remains are identified with part of the southern wall of the nave. The oldest part of the temple is undoubtedly the main apse, an **11th century** work, with both side apses, the tower and the el portico corresponding to subsequent interventions undertaken between the end of the 12th century and the early 13th century. Since the Ecclesiastical Confiscations of 1835 it lost its status as a parish church and became answerable to La Trinidad[82], this marking the start of its decline and being subject to numerous uses until in the early 20th century it was bought by the famed potter

CHRONOLOGY OF ST. JOHN OF THE KNIGHTS

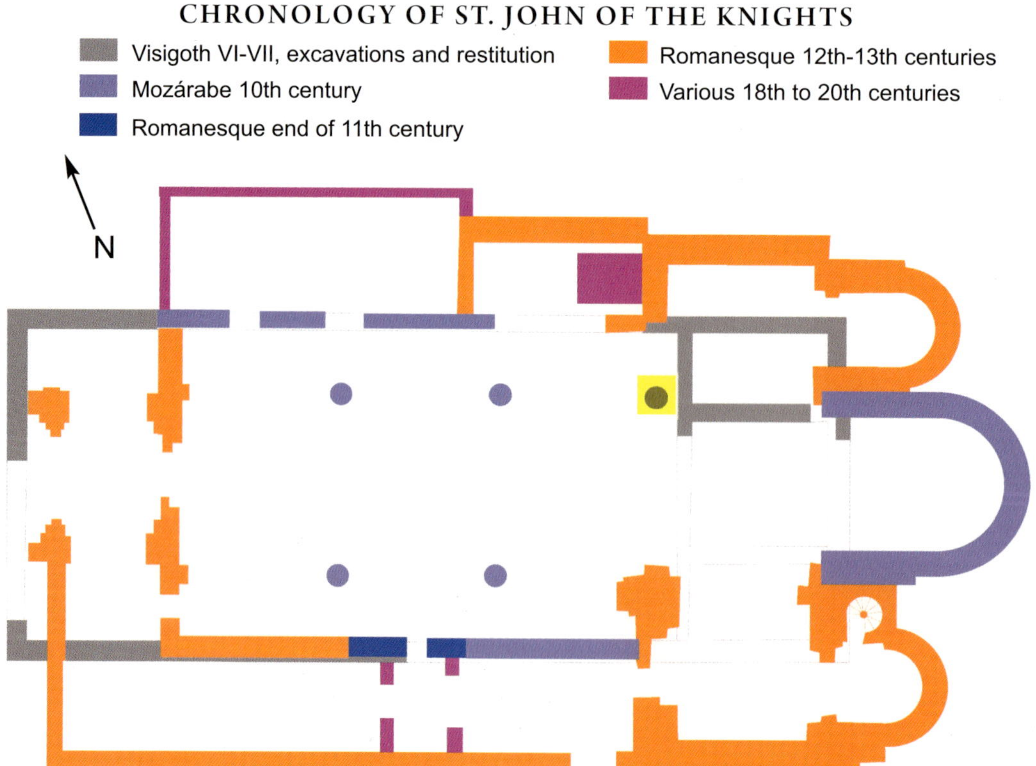

ST. JOHN OF THE KNIGHTS 52

San Juan de los Caballeros

Daniel Zuloaga who converted it into his workshop and dwelling; years after the artist's death the temple was acquired by the State, setting up a museum there dedicated to the work of its former owner.
Worthy of special mention on the exterior is its **porticoed gallery**, closed off in the 17th century and subsequently recovered, comprising dual columns which support beautifully made dual spires, but hard to read because of their advanced state of deterioration owing to erosion; even so, two scenes can be made out, one of the Annunciation and the image of God the Father. The eaves are significant too, comprising richly decorated **modillions** and **metopes** inside small trilobite arches. On the westside, the main gateway opens out, decorated in the style of Ávila and proceeded by a narthex of Gothic-like appearance constructed in the 13th century.
Inside we are presented with a simple, transparent space where the sculptural decoration is limited to the transept whose spires denote Burgos influence. Remains of the pictorial decoration which embellished this area are still visible but, as with the spires, they are highly deteriorated. What's more, in the access to the main apse a cancel is raised with the shields of the Contreras who had their pantheon here.

Western gateways

ZULOAGA MUSEUM

In around 1905 the church of St. John of the Knights was acquired by the potter **Daniel Zuloaga** who set up his workshop and house there. Trained in **Sèvres**, he carried out part of his artistic activity in Segovia along with his children and his nephew the painter **Ignacio Zuloaga**. During the years he worked at the church he gradually adapted it, installing the kilns and the workshop in the naves and in the upper part of the latter putting up a structure where his dwelling was located; the portico was used as a room for meetings and for displaying and selling works and the tower was set up as a painting studio. After the artist's death, the temple was acquired by the State and a museum was set up there dedicated to his work; various ceramic pieces were kept there as well as drawings and sketches and antiques that he liked to collect and sell. At present his great grandson is continuing his work with a workshop situated in Merced square, Segovia.
Zuloaga's artistic production, continued through his children, is characterised by the **recovery of traditional** Spanish techniques, with Hispano-Moslem, Triana and Talavera of the Baroque serving as a reference; as well as this look back towards the past, he also introduced **new fashion styles** in Europe and new techniques. However, **regionalism** was the prevailing note in his works by means of the implementation of Castilian landscapes and types which has been what has defined his work most.

Pottery kiln at a chapel of the church.

RUINS OF ST. AGUSTINE'S CONVENT [51]

All that remains of the old convent are the transept facings and the crossvault of its church as the naves, the roofs and the rest of the outbuildings have gradually disappeared over the centuries to give way to other constructions. It was a building endowed with large dimensions constructed in the mid-16th century and it is currently used as an open-air music auditorium. Daniel Zuloaga was involved in the struggle for its survival and he wrote articles strongly criticising its demolition, but he was not able to prevent it from coming down.

PROVINCIAL COUNCIL PALACE [53]

Façade of the Uceda-Peralta Palace

This institution has its head offices in two former palaces which have been totally overhauled in order to adapt to the new administrative uses. The former is known as the Uceda Peralta Palace; a 17th century work with a simple façade on which its classicist gateway stands out.

The latter is the Palace of the Marquis of Castellanos, also known as the Maldonado, dating back to the 15th century and, despite the transformations, still preserves the original remains in its façade such as the lintelled gateway framed by an alfiz decorated with balls which houses the coat-of-arms of the Maldonado.

Gateway of the Castellanos Palace

QUINTANAR PALACE 56

Also known as the Casa de las Cabezas (House of the Heads) owing to the decoration shown on its gateway; its construction dates back to the late 15th century and the early 16th century, though in its interior courtyard the remains have been found of a late Romanesque gateway. The most significant thing about the building is its gateway adorned by helmets and framed by an alfiz which houses two wildcats whose carving work is quite rough, holding the coat-of-arms of the family. In recent restorations some interesting murals have been found in its interior with motifs of grotesques dating back to the late 18th century.

COUNT OF CHESTE PALACE 54

This 15th century building was originally owned by the alderman Juan de Contreras, subsequently belonging to the Count of Cheste in the mid-19th century; it currently houses a school run by the Conceptionist Mother Superiors and its internal structure has been changed greatly to this end.

On the exterior the semicircular archway has been preserved, framed by a simple broken alfiz which accommodates a 17th century coat-of-arms belonging to the Contreras lineage; more or less taking up the central axis of the façade the remains of the main gateway of the palace are preserved, now closed off, and a mullioned window endowed with fine Gothic tracery.

In the interior there is a small, very simple courtyard with shields on the capitals of the columns which support the architraved structure.

It is worth pointing out that during the 19th century and the early 20th century this palace housed an important literary circle chaired by the Count of Cheste himself and assisted by the young Marquis of Lozoya.

MARQUIS OF LOZOYA PALACE[55]

Located alongside the now disappeared St. John's gateway, this house formed, along with the Palace of the Marquis of Moya, part of the city's defences; its construction dates back to the 13th century and it is known that there was a Romanesque tower alongside the city walls which strengthened surveillance at the gate. This tower disappeared as the same time as the city wall gate in the 19th century. From this Romanesque period other remains have been preserved in the interior besides the main gateway, richly decorated by rosettes, belonging to a gateway and some wooden coffered ceilings with decorative motifs.
During the course of time it has undergone various remodelling works and been in the hands of different owners, belonging to the Eldest Son Entailed estate Mayorazgo de los Cáceres and at present to the Marquis of Lozoya.

HOUSE OF CHAINS or SEGOVIA HOUSE[57]

Also known as the House of Chains, this belonged to the Marquis of Moya, Mr. Andrés Cabrera and Mrs. Beatriz de Bobadilla, a close friend of Queen Isabel the Catholic since childhood. This fortress-like house formed part, as did Casa de los Picos[20], of the city's defence system, strengthening - in this case, along with the house of the Marquis Lozoya - the old gateway of St. John. This defensive nature of the house dates back to the 13th century, at which time there was already a construction adjoining the city walls and whereof some vestiges have been kept such as the circular tower alongside the church of San Sebastián. However, the building is essentially a 15th century work with a large access gateway crowned by the arms of their illustrious owners and organized around a central courtyard decorated by Isabeline balls.

The house of Segovia from the Azoquejo

ST. SEBASTIAN OF THE KNIGHTS[58]

The building is situated in the district of Caballeros, near the end of the aqueduct[12]; over the course of time it has suffered numerous transformations which have affected its internal appearance, with the exterior remaining in the best state of repair. The temple as built in the **12th century**, comprising a nave, a **semicircular apse**, a tower and a portico; the porticoed gallery has disappears as a consequence of the remodelling work and the sanctuary is the area in the best state of repair. This stands out for its prolific **sculptural decoration**, located in the apse windows and in the cantilevers; both the iconographic repertoire as well as the manufacture are connected with a workshop which worked on other churches in the city such as the one near St. John of the Knights[88] or St.Milan[114] and St.Martin[32] or influenced by the Romanesque works of the city of Ávila, with its main reference being the basilica of Saint Vincent. The **gateway** has also been preserved from the original manufacture whose keystones were decorated by a fine carving with vegetal motifs.

Apse and small tower of Casa de las Cadenas

Detail of the gateway

FORMER COLEGIO DE LA COMPAÑÍA[61]

The works on this major classicist temple were started in **1577**, following the designs by the Jesuit Valeriani and corrected by **Juan de Herrera**; however, the construction soon stopped, being resumed in 1582 under the orders of another Jesuit Andrés Ruiz. The figure of Juan de Herrera was present throughout the construction process and it could be said that he was its intellectual author; by contrast, the site managers came and went, with some as noteworthy as **Pedro de Brizuela**. The construction model for tis building was the **Collegiate Church of San Luis de Villagarcía de Campos** in the province of Valladolid which it follows faithfully and its façade is a replica of the former, except for the ashlar whose profile is slightly highlighted here. Also like the former, it has a Jesuitic Latin-cross layout contained within a rectangle with chapels on the sides of the nave, a crossvault covered by a semi-spherical vault and lighting with spans.

The school erected alongside the church was also built in line with the standards of the purest classicist Baroque style.

After expulsion of the Jesuits, the whole building was used as a Seminary.

CONVENT OF THE IMMACULATE CONCEPTION[59]

The Franciscan Conceptionists are currently accommodated in an old Baroque palace near the end of the aqueduct with its presence in the city dating back to the early 17th century. The Order of the Immaculate Conception was founded in the mid-15th century by **Santa Beatriz de Silva**, the first lady of Queen Isabel, the second wife of Juan II and the mother of Isabel the Catholic; said lady had a vision of the Virgin for three days when the queen was shut in a trunk because of jealousy, ordering her to found a new Order in her honour with the same white and blue habit in which the Virgin had appeared to her. Nuns are renowned for their high-quality **confectionery**; this is so successful that two of them have published several books of recipes and are the stars of a TV cooking programme.

LOS AGUILAR PALACE[60]

This is one of the best Baroque **façades** in Segovia; the building was acquired in the early 17th century by the Aguilar family, lords of Encinas, who obtained the title of counts in the early 18th century, the period from which the beautiful coats-of-arms decorating the gateway date.

VA SE

A Q

E SEGOVIA

O V I E N S I S

ALTO DEL CALVARIO

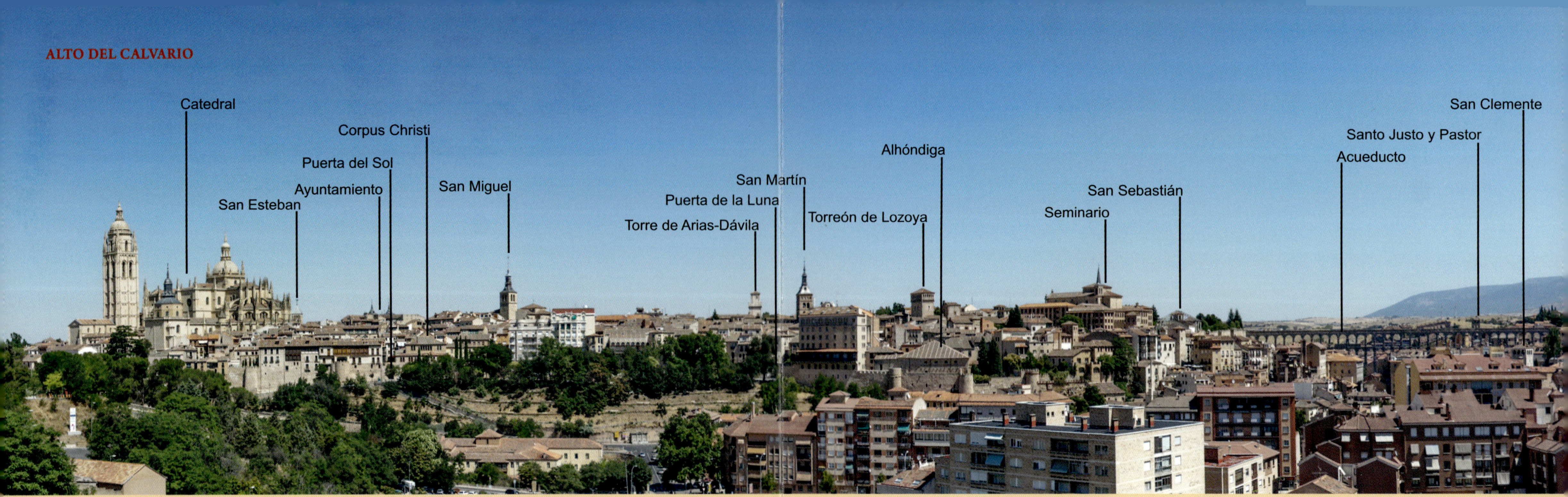

Catedral
San Esteban
Ayuntamiento
Puerta del Sol
Corpus Christi
San Miguel
Torre de Arias-Dávila
Puerta de la Luna
San Martín
Torreón de Lozoya
Alhóndiga
Seminario
San Sebastián
Acueducto
Santo Justo y Pastor
San Clemente

SEGOVIA

MIRADOR DE EL TERMINILLO

Santos Justo y Pastor
San Lorenzo
Acueducto
Casa de Segovia
San Sebastián
San Juan de los Caballeros
Seminario
San Agustín
San Martín
Torre de Arias-Dávila
Santa Cruz la Real
San Miguel
La Trinidad
San Nicolás
Torre de Hércules
Ayuntamiento
Catedral
San Quirce
San Esteban
San Vicente el Real
San Andrés
San Pedro de los Picos
Puerta de Santiago
Alcázar

CHURCH OF SAINTS JUSTO AND PASTOR[62]

This **Romanesque** church is located in the clothmakers' district and it was constructed between the **12th and 13th centuries**; like many others, it has legendary origins connected with the devotional image of **Christ of the Gascons**. This legend has it that the image was disputed by Gascons and Germans who decided to put it on a mule and build a temple in the place chosen by the animal, in such a way that the beast dropped down dead in the plot today occupied by the church. Owing to the devotional nature of the image, the church has suffered numerous extensions over the centuries which have configured its current appearance; the oldest elements are the sanctuary, the nave and the tower, to which a boardroom was added in the 17th century for the brotherhood of the Holy Slavery of Christ of the Gascons. On the exterior its **tower** stands out, similar in appearance to the nearby church of El Salvador[100];

the capitals adorning its windows reveal influences from the Basilica of San Vicente de Ávila. In the interior, the Access to the tower, discovered in the 1960's, has a beautiful, richly adorned **little door** whose **tympanum** depicts a scene linked to the discovery of the Holy Cross by Saint Helen in which the latter can be seen in the centre of the composition dressed in rich attire and crowned, accompanied by two ladies and San Macario, the bishop of Jerusalem and an angel purifying the altar with the Cross. The depiction is very well made and its carving is beautiful and painstaking, similar to the Gothic style, yet still Romanesque. The ground floor of the tower must have housed the chapel of Christ of the Gascons which was extended in successive remodelling works until arriving at its current state. Its interior harbours the image of **Christ of the Gascons**, **Romanesque** from the second half of the **12th century** made of polychromate wood which has the peculiarity of having arms articulated at the height of the shoulders and elbows which would allow him to be shown as crucified or supine in the different liturgical portrayals of Holy Week. As with all Romanesque Christs, it is characterised by simple, schematic treatment of the anatomic features and pronounced rigidity. The Christ is taken out on a procession every Thursday and Good Friday by the Royal Brotherhood of the Holy, Venerable Slavery and of the Holy Burial of Christ of the Gascons.

This church also conceals one of the most significant examples of Romanesque **mural painting**; discoveries in the second half of the 20th century they were developed by the Marquis of Lozoya who saw to their dissemination. They are situated in the sanctuary and they incorporate a varied iconographic repertoire, highlighting the portrayal of the Lord's Supper with the inclusion of St. Paul and the Arrest scene; presiding over proceedings is Christ in His Majesty in a mandorla in the sacrament niche of the apse. The Marquis of Lozoya saw the work of two workshops or different masters which differ in the postural treatment of the figures and in the relationship of some characters with others, resulting in those of the master, older, colder with more rigid figures and with those of the second being characterized by their dynamism and the search for a certain perspective. These paintings are clearly influenced by the Devout of Tábara, books illustrated in the 10th century in its famed Scriptorium.

ICONOGRAPHIC PROGRAMME

1 Elephant with tub
2 Original Sin
3 Bestiary
4 Cain and Abel
5 Agnus Dei
6 Saints and Martyrs
7 Indetermine scene
8 The Arrest
9 The Lord's Supper
10 Soldiers
11 Maiestas Domini
12 The 24 Elders of the Apocalypse

The Tetramorphs

13 St. Matthew
14 St. John the Baptist
15 St. Luke
16 St. Mark
17 The Crucifixion
18 The Descent
19 The Transfiguration of Christ or Doubting Thomas
20 Old holes to hang the articulated Christ during the Holy Week liturgy.

THE SAVIOUR[63]

This former parish church of the **clothmakers' district** was built near the aqueduct[12] where the galleries can still be seen which crowned the dwellings and which were used as drying places. The church is a **Romanesque** building from the late **12th century** and early 13th century which shares similarities with the other church in the district dedicated to Saints Justo and Pastor; these similarities can be seen from the exterior appearance of its **tower** despite the fact it was submitted to notable modifications during the 16th century. From the Romanesque age, in addition to the lower floors of the tower, it preserves the gateway and the **portico**, with the latter being the most noteworthy; it currently preserves five arches on dual **capitals** which show very varied iconography, though some are highly deteriorated, making them hard to read; the historiated capitals bring together the scenes of the Adoration of the shepherds, the Epiphany and the Massacre of the Innocents. The high part of the portico is surrounded by intricately decorated **metopes**, the majority being based on vegetal and geometric motifs.

In the **16th century** the church underwent major modifications and the sanctuary polygonal comes from this time, designed by **Rodrigo Gil de Hontañón** following the standards of the time with elements which are still reminiscent of the Gothic.

THE BERMEJO HOUSE[64]

Very near the start of the aqueduct[12] bridge you will find this **15th century** house, a model of civil architecture of what was known as **Elizabethan Gothic**; it has a façade with a lintelled gateway framed by a broken alfiz decorated with balls, a decoration typical of the age of the Catholic Monarchs, and crowned by the coat-of-arms of the former owners, being one of the best carried out and preserved in the city. On the second floor two windows open out which are also framed by decoration with balls. The rest of the building has undergone many modifications over the course of time, not retaining its original structure.

Chapel of the Convent of Saint Isabel

Grille crowning

SAINT ISABEL CONVENT 65

The former Saint Clara la Vieja Convent, the nuns of Saint Isabel arrived here at the behest of Queen **Isabel the Catholic** in the late 15th century, though the foundation of the Order dates back to 1486. The convent **church** dates back to the **mid-16th century**, but it still has elements characteristic of late Gothic. In its interior you will find the oldest **Plateresque grille** in Spain carried out in 1507 for the former Cathedral with influences from the Burgos school.

Today the community of nuns prepares a wide variety of exquisite sweets.

INCARNATION CONVENT 66

Very near the Casa del Agua (Water house) or sand remover of San Gabriel, where the first arches of the aqueduct were raised, the monastery of Augustine Mother Superiors was constructed, called Saint Rita or Humble Incarnation, a 16th century work.

Church of the Incarnation Convent

ST. ANTHONY'S EL REAL MONASTERY

HISTORY

This was founded in 1455 in the vicinity of the Aqueduct sponsored by King **Enrique IV**, at a recreational estate possessed by the latter on the outskirts of Segovia, called El Campillo, a convent for Observant Franciscans; at the outset the Franciscans settled in some outbuildings known as the **House of the Prince** whilst they were building the new convent. This comprised two spaces, one rectangular and the other quadrangular, double-height and imbued with great decorative wealth.

In 1488 **Isabel I** the Catholic granted the site to the Poor Clares who undertook remodelling work to adapt the space to their convent needs, adding, for example, the choir stall at the foot of the church.

The importance this convent eventually had is borne out by the **three Mercedes** of the aqueduct with which it was endowed, having discovered two records of one of the Mercedes during recent remodelling work at the Friars Room.

It is believed that Enrique IV may have thought about being buried in the monastery, the same as his sister, having left this written in his will, clarifying that he wanted his burial to be temporary.

The richness of this convent can be seen from its good state of repair as it has managed to remain until today in the same condition as when it was built, keeping in its interior a repertoire of high quality, very beautiful **coffered Mudejar ceilings** which do not have a comparison.

ENRIQUE IV

Born in Valladolid in 1425, at a tender age he moved to Segovia, the city with which he had a close relationship and which was one of his favourites along with Madrid. The son of Juan II and María of Aragón, his father granted him the feudal estate of Segovia in 1440 on the grounds of his marriage to Blanca de Navarra, who he separated from after some years owing to the inability to produce a descendent. In 1455 when he was already King, he married Juana de Avis who bore him a daughter known as Juana la Beltraneja, not recognised as the king's daughter by a faction of the nobles and which led to a civil war after the sovereign's death.

History has shown him to be a weak, sickly man whose character was fickle and capricious; however, he was extremely cultured and had a particular interest in the arts and his friends included Garcilaso de la Vega, and the promoter of major works where the setting for his action was the city of Segovia with which he had maintained a very special relationship since his childhood and which became one of the most important power centres of his kingdom. In this regard, he undertook works at the Alcázar, founding the monastery of San Antonio el Real at his recreational estate and expanding his Royal Palace; in all of them his lining for the oriental stood out, using Mudejar builders as workers who left as a testimony to this age the most beautiful coffered ceilings of the 15th century.

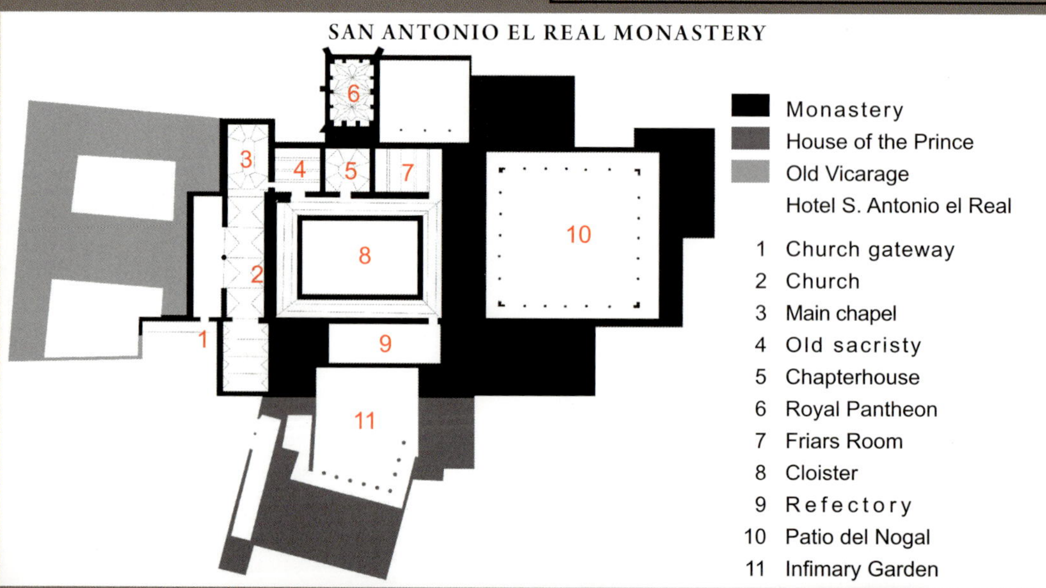

The **church gateway** is attributed to the circle of **Juan Guas** and **Enrique Egas** by dint of its composition based on superimposed structures framed by an alfiz in whose interior two coats-of-arms of Enrique IV are depicted.

On the exterior, the convent **façade**, belonging to the **Casa del Príncipe** (Prince's House) is endowed with a gallery whose upper part is open with Tuscan arches which are also used in the lower storey access. On the one side another door can be seen, carried out in the 18th century, where, sheltered by niches, there is a depiction of Enrique IV accompanied by St. Anthony of Padua and his sister Isabel accompanied by Santa Clara; both monarchs are in praying position. The whole is rounded off by a royal coat-of-arms and on the sides two Franciscan coats-of-arms.

The **church** has sober, austere appearance in contrast with the richly decorated main chapel which is why for a long time it was believed that this was the prince's house. The church was remodelled in the 18th century, but its splendid **15th century Mudejar coffered ceiling** was respected, carried out with Valsaín[152] pine wood; this is an octagonal armour richly decorated by five and ten-point stars, polychromed in blue, red and gold and whose symbolic purpose was the royal exaltation on the grounds of the victory of Enrique IV over the Moslems at Jimena; to this symbolism there accrues the metaphoric celestial representation. The frieze running underneath is delimited by the Franciscan cord and the coat-of-arms of Enrique IV is repeated there, surrounded by pomegranate tree branches of grenades and Gothic motifs. In the nave you will find the **Retable of the Passion** carried out in the second quarter of the **15th century** by a Flemish artist and imported to Spain from the Netherlands. Despite having suffered remodelling work and repainting in the 18th century when it was moved to its current site from the main chapel, it can be seen in all its splendour; it has been carried out in carved, polychromated wood and a set of assembled blocks. It is endowed with **more than 100 round-shaped figures** which are characterised by their expressiveness and by the individualization of their gestures and faces, making them wonderfully expressive and imbuing the whole with a theatrical nature. The retable has been carried out with a pictorial conception, endowed with clear references to the works of **Jan van Eyck** and **Roger van der Weyden** both in terms of the attitude of the characters as well as their clothing and placement. The theme of the Passion is portrayed through various secondary stages which occur around the central theme of the Crucifixion, making an odd use of perspective; the central theme is complemented by some small secondary groups located on the sides of the retable.

Retable of the Passion

Mudejar coffered ceiling

From the church access is gained to the **sacristy** covered by a beautiful **polychromated wooden ceiling** decorated by vegetal motifs and the coats-of-arms of Enrique IV and the Franciscan cord.

From here the **Main Cloister** is reached whose passages are covered by coffered ceilings which are simpler than those seen up to now, but perfectly preserved as the holes of the cloister were closed in the 18th century. As with a museum there is a variety of objects on display, highlighting the collection of images of the **Baby Jesus** and the three **Flemish triptychs by Utrech** in which painting and sculpture are combined; what is noteworthy in them is the central scene in high relief made of polychromed earthenware with the themes of the Calvary, Christ with the cross on his back and the Holy Burial, whilst the sides are boards painted with figures of saints.

Detail of the coffered ceiling of the cloister

The most important rooms of the convent open out onto the cloister; the **Refectory**, a large rectangular room around which there is a bench where the nuns would get together to eat; its walls are covered by murals with saints and vegetation, highlighting the paintings of the top with portrayals of Christ, Saint Clare and the Immaculate Conception. Reference should also be made to the pulpit where the readings were given which is profusely decorated by vegetal motifs and which dates back to the 15th century.

This is followed by the **Friars Room** whose ceiling boasts a decoration similar to that of the sacristy. In this room two water fountains were recently found which connect to a channel deriving from one of the Mercedes of the aqueduct.

Retable of The Calvary

North side of the cloister

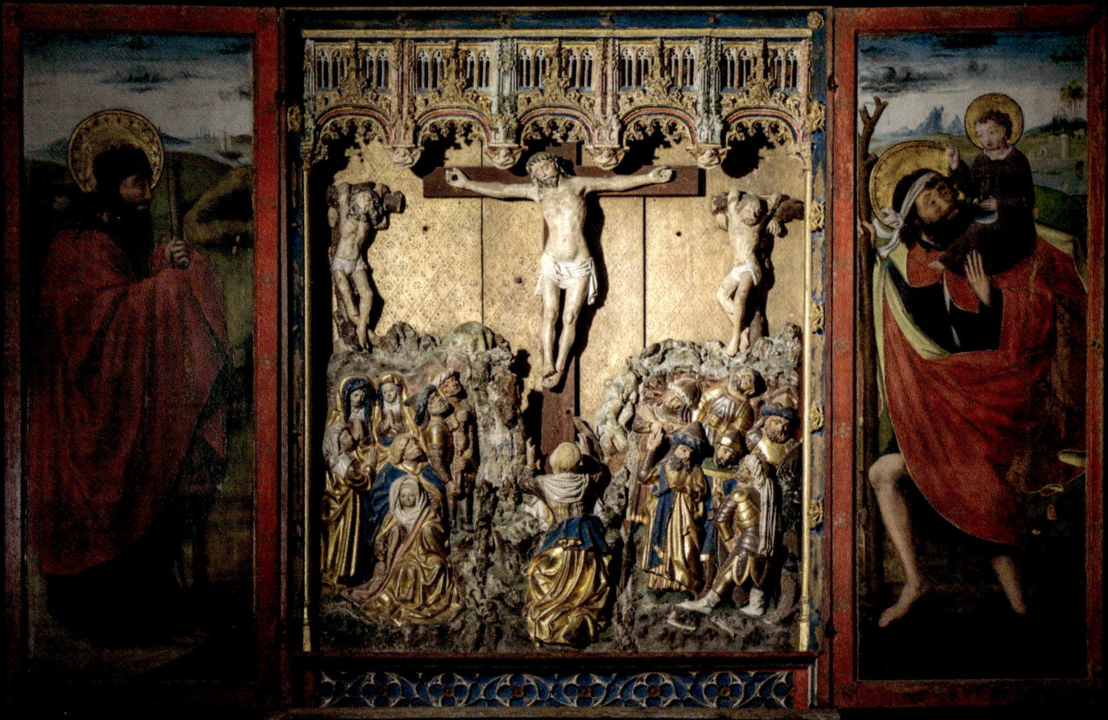

Mocárabe

Finally, the **Chapterhouse** covered by a spectacular octagonal **coffered ceiling** polychromed in blue, red and gold and decorated by the coat-of-arms of Castile and Portugal together with Franciscans symbols; his vision is impressive, even more accentuated by the low height at which it is situated.

The coat-of-arms of Castile

Coffered ceiling in the Chapterhouse

St. Eulalia square constituted one of the best examples of **16th century civil architecture** with its blazoned houses and its supports with granite columns and wooden latticework in the more modest ones; however, its image has changed over time and the Clamores stream which crossed it has been channelled and some dwellings have been knocked down, only partially retaining the supports with their heraldic coats-of-arms and the **house of the comunero Antonio de Buitrago** which boasts a gateway famed by a ball-bedecked and a shield as a result of the punishment dealt out by Carlos I to those who had taken part in the Revolt of the Comuneros

SAINT EULALIA[68]

In the centre of the square there stands the church from the early **13th century** from where it gets its name; also subject to many transformations in the 17th century with extensions and Baroque remodelling work, it still retains part of its **Romanesque** past. Worthy of special note in the exterior are its truncated topped **tower** which has certain similarities to that of the church of Saint Justo[97], its **southern gateway** and the **sculptural decoration** of the interior.

SAINT THOMAS[69]

Located outside the walls this is one of the latest Romanesque parish churches in the city built between the late **12th century** and the early 13th century. Of its medieval past there only its **semicircular apse** and the **northern gateway** remain in good condition and the rest of the building was subjected to intense remodelling work during the 17th century, very evident in the interior whose vaults have been covered by Baroque plasterwork.

ARTILLERY ACADEMY 70

The current Artillery Academy takes up the plot of the **former convent of Saint Francis** whose monks arrived there after abandoning the monastery of San Antonio el Real[102]. All that remains of that convent are the double-height **courtyard** from the **15th century**, with the lower storey comprising segmental arches and the upper storey made up of trilobed arches and beautiful tracery in the parapets; owing to its characteristic it is attributed to **Juan Guas**. Its outbuildings house the Artillery Museum. The Artillery Academy was founded in 1764 and it is **the oldest one in the world which is still active**, arriving here in 1863 from the Alcázar[56] which it had to abandon as a result of the fire which devastated it in 1862.

CASA DEL SELLO 71

During the 15th and 16th centuries there was a major textile industry in the city of Segovia and the cloths produced were highly appreciated for their quality; however, in the 17th century the industry took a downturn, partly brought about by the fall in the quality of the fabrics, leading to the creation by order of the Castile Council of the **Clothmaking Regulating Council** whose main purpose was to ensure good practices and the quality of the cloths, providing them with a seal to assure this, somewhat similar to the work carried out today by the Denominations of Origin. This institution was housed in a building in the late **16th century** which stands out for its **façade** that has retained its original appearance as the interior has undergone many transformations. Of particular note are the thick, classical-style columns framing the façade; the rest is sober and well made.

ST. CLEMENT 72

Situated on the banks of the former bank of Clamores you will find this **Romanesque** church dating back to the early **13th century**. It has a nave finished by a semicircular apse where the sculptural work is concentrated, highlighting the finishes of the pilasters which divide and liven up the apse panels; said finishes are formally related with the cimborrio of **Zamora Cathedral**. The temple also comprises a tower, two gateways and a porticoed gallery which has similarities with that of La Trinidad[82]. The temple was subjected to major restoration work in the mid- 20th century, providing it with its current appearance and removing all the exterior additions; however, this was not the only intervention that the church has been subject to as in the 18th century its interior was redone and covered in Baroque plasterwork; only the southern **apse chapel** retains its original Romanesque appearance, reminiscent of the chapels of this type to be found in the churches of San Quirce[80] and San Nicolás[81]. Both the shell of the apse as well as the presbytery path are adorned by **murals** from the late 13th century, already imbused with the **linear Gothic**, though reminiscences of the Romanesque are still perceptible. In the apse there is a depiction of **Maiestas Domini** inside a mandorla flanked by the Tetramorphs and two seraphins; under them there are six characters carrying building

Scales of the apse of Saint Clemente and the cupola of Zamora Cathedral

blueprints who have been identified as the **Doctors of the Church**. On the presbytery paths there is a depiction of the Genealogy of Christ with Tree of Jesse, from whose trunk there are scrolls with the predecessors of Jesus and which starts in **Jesse's belly**, ending in a portrayal of the Virgin with a representation of Baby Jesus; the iconographic programme of these paintings alludes to the pillars of Christianity and of the church.

CASA DE LOS AYALA~BERGANZA$_{74}$

This late **15th century** building is known in Segovia as the **House of Crime** as in 1892 the owner of the dwelling and a servant were killed by thieves in its interior; ten years later, this house became the workshop of the great Spanish painter **Ignacio Zuloaga** who occupied it with his friend and also artist, Pablo Uranga. It was the latter's testimony that added to the dark legend of this building as he said he had seen an aquelarre of witches in the basement and this narrating seems to have served as the source of inspiration for the famous painting *The Witches of Saint Milan* by Zuloaga.

The building is currently perfectly restored, retaining those elements characteristic of its construction age; hence, worthy of special mention is the high gallery formed by ogee arches of brick resting on convex columns and granite capitals and which is finished off by a cornice with gargoyles made of the same material. The main façade still retains its original structure, highlighting the gateway and a window with a granite frame. The decorative solutions appearing in this house are already framed within the **Plateresque style**, the precursor of the Renaissance.

CASA DE LA TIERRA$_{73}$

This was built in the **mid- 18th century** by the royal project manager José de la Calle on the plot of the former Pueblos House from the 15th century; this belonged, and still does, to the Comunidad de Ciudad y Tierra comprising the sexmos (medieval administrative division) representatives whose used it as a chamber and meeting place, also housing the royal entourage when the monarchs visited the city. The building is structured around a courtyard and it boasts a simple **façade** adorned by paintings portraying human figures in a poor state of conservation, making it hard to interpret. The door is finished off by a **medallion** bearing a winged figure with a very simple design which could well be a depiction of Mercury, the God of travellers, of the orders and of cattle.

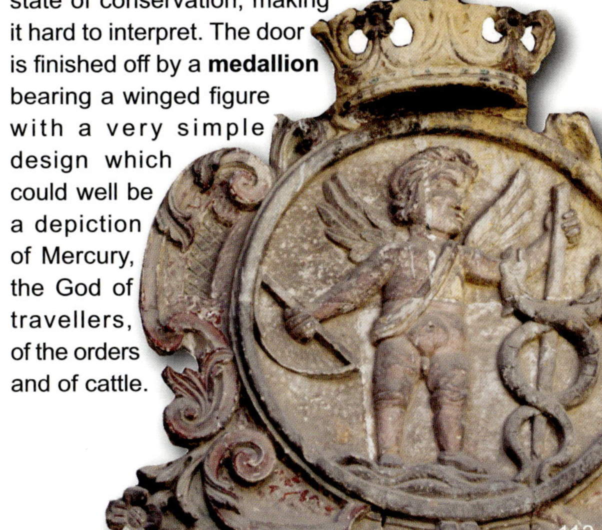

SAN MILLÁN[75]

This is the largest, most significant Romanesque church in the city. Constructed on a **Mozárabe necropolis** and an alleged temple from the same of which no trace remains except for the current **tower** which, by dint of the materials used in its construction, lime, and the presence of horseshoe arches in its windows, could well be the bell tower of the former construction. The current Romanesque temple was built during several construction campaigns, having been started in the first half of the **12th century**; by observing its plan great similarities can be seen with the **Jaca cathedral** with which it also shares the alternance of cruciform pillars with adjoining semicolumns and columns. This similarity has been justified by the presence during the first half of the 12th century of an Aragonese population in the area as the result of the influence and dominion that the Aragonese monarch Alfonso I the Warrior wished to exert in this part of the kingdom, the husband of Queen Urraca of Castile with whom he had major territorial disputes.

The church boasts three naves with a high transept and four apses; the body of the church with the three main apses belongs to an initial construction stage; the construction of the fourth apse between the tower and the apse of the Gospel and the two porticoed galleries correspond to a later stage in the 13th century.

In terms of its construction and structure the temple not only has jaquesas influences, but also of note is the contribution by Moslem master builders who in all likelihood planned the dome, and, furthermore, the influences deriving from major temples built on the edge of Way of St. James. Sculpturally speaking, the reference to the **basilica of San Vicente de Ávila** is continuous, observing the work of a master trained in the workshop of Ávila who, with less skill, conveyed here what had been done there.

Southern gateway

On the **exterior** the church looks round, though it has lost its purpose as a centre bringing together one of the densest populated suburbs in the city, brought about by the major urbanistic remodelling to which the area was submitted and the channelling of the River Clamores which ran in the vicinity of the temple; even so, it looks beautiful, highlighting its porticoed galleries supported on

CHRONOLOGY OF THE CONSTRUCTION

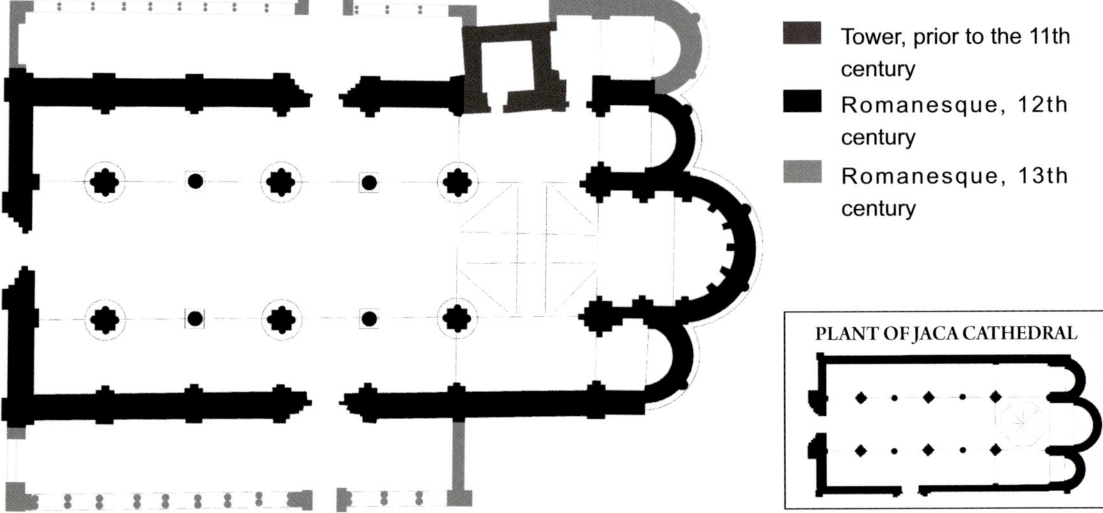

double columns. The **southern portico** boasts richly carved arches on which some historiated scenes can be observed with themes like the Annunciation, the Visitation, Epiphany, a Pantocrator surrounded by the Tetramorphs or the Arrest of Christ. In turn, the northern portico underwent major intervention work in the 16th century. In both the influence of the workshops of Ávila can be seen and the similarities with the churches of Sepúlveda[154] and Fuentidueña[154].

As far as the **gateways** are concerned, the most noteworthy is the western one whose layout is reminiscent of those of the churches of San Esteban and Santiago del Burgo in **Zamora**, including the central body of the Bishop's Gate at the Zamoran Cathedral. Here there is clear evidence of the change in design that was adopted to cover the naves which were initially to be vaulted and which ended up being covered by a **wooden armour** of which there are still some remains salvaged from a restoration and which were concealed by the Baroque vaults which were set up in the 17th century; it constitutes a major example of Hispano-moslem art in the 12th century adorned by delicate knotwork and kufic inscriptions.

In the **interior**, in addition to its clear references to the Jaca cathedral, there are significant influences by San Vicente de Ávila; hence, in the main chapel, decorated by arches, we can find capitals with clear references to the Ávila temple such as that portraying a knight and a lady, finely carved, but less detailed than its reference. Also of particular note are the **capitals** adorning the pillars of the naves on which there are again continuous references to the Ávila temple both in terms of the treatment of the carving and the themes; the castle or elephant theme are repeated. Particular attention should be paid to that showing the scene of the **Flight into Egypt**, carried out in all kinds of details and the one representing the **Epiphany** whose composition is detailed.

ST. LAWRENCE'S SQUARE

The district of San Lorenzo has been able to preserve like no other the essence of traditional medieval-based architecture; it is the expression of the rural environment in the city as its settlers were generally involved in farm work in the orchards and mills on the banks of the Eresma. The houses

reflect this way of life; they comprise two storeys and on the exterior they reveal their wooden structure in the framework with brick and mud, sometimes joined by ashlar gateways and the remains of columns and other stony elements.

ST. LAWRENCE[76]

In the centre of the square there lies its parish church whose chronology is indeterminate and whose structural elements, from between the **12th and 13th centuries**, though some people say that they derive from an earlier period in the 10th and 11th centuries and even the 9th century, being based on the remains of a semicircular sanctuary which appeared during some restoration works and which demonstrate the existence of a previous temple on the same plot. However, it should be pointed out that the excavated remains are Romanesque and they belonged to a temple probably from the early 12th century, endowed with a simple design which was replaced over time. The oldest part of the church is its nave with the two gateways, with the dome head of a triple apse and the tower belonging to a later extension stage. There are numerous **archaic** decorative elements it boasts and which have given rise to the possibilities that link it to a Mozárabe building, though they are no more than reinterpretations of elements from other eras used in the Romanesque period and which are reminiscent of the Zamoran Romanesque in which this type of device is very common.

The church stands out for the sculpture of the chapters, both of the windows of the apse and of the portico; in the former, scenes are depicted related with the **martyrdom of San Lorenzo** such as the decapitation of Pope Sixtus or San Lorenzo's own ordeal, culminating in death as well as Isaac's sacrifice.

The portico was built on paired columns with chapters whose theme is greatly influenced by the portico of the Duratón church and which is also present in other temples of the city such as St. Martin[32] and St. Milan[114]. Generally speaking, the state of repair of the chapters is not good and this makes them hard to read; even so, some scenes can be made out such as the Adoration of the Magi or the Massacre of the Innocents, along with others depicting animals and men fighting.

Martyrdom of San Lorenzo.

Isaac's Sacrifice

ST. VICENT EL REAL CONVENT 77

Its origin is old and imprecise and legend would have it that it is linked to the existence of a Roman temple and later to a hermitage on the same plot. The current monastic complex was constructed during three construction campaigns which overlaid the remains of the previous ones, giving rise to a whole which it is hard to read. It is known that in the mid-12th century a Cistercian monks' foundation was set up in the capital, the third in the province, who erected their house here; this construction would seem to date back to the early **13th century** and almost nothing remains of it except for some Romanesque vestiges and part of the church as in the s.14th century it suffered a fire which left it greatly damaged, undertaking renovation work during the 15th century to replace it. In the early 17th century the convent suffered yet another fire after which **Pedro de Brizuela** took charge of the renovation and current appearance in Baroque style.

SANTA CRUZ LA REAL CONVENT[78]

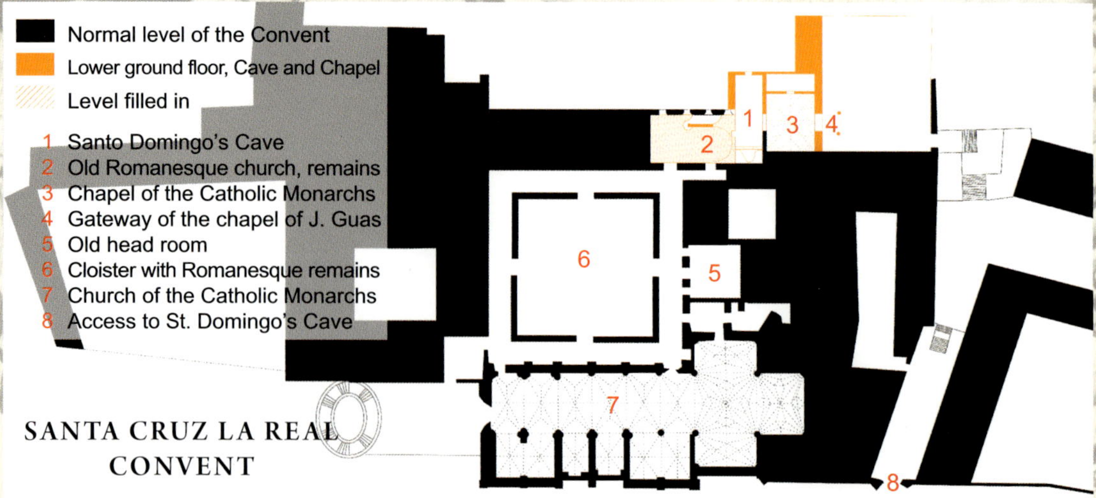

Near the stream of the River Eresma there lies the fir**st Dominican foundation in Spain**; its origins date back to **1218** whilst **Santo Domingo de Guzmán** was staying in Segovia, only two years after the foundation of the Order of Preachers. It reached its height in the **15th century** during the reign of the **Catholic Monarchs** who decided to empower the Order and granting it inquisitorial work. During these years, in line with the wishes of Queen Isabel, the convent was expanded and a new chapel was built with just one nave following the Hispano-Flemish style in vogue at the time and whose greatest exponent was **Juan Guas**, its author and of the chapel of the monastery of El Parral[130], in addition to the cloister[50] of the former cathedral. Prior to the remodelling work by the Catholic Monarchs there were a series of outbuildings which made up of the house of the monks, with numerous Romanesque remains still visible and integrate into the current architecture. Worthy of

Gateway to the Santo Domingo cave

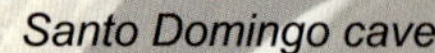

Santo Domingo cave

ie

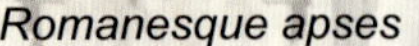

Romanesque apses

Romanesque gateway

Romanesque Span

special mention is the former Romanesque chapel, erected at a lower level and filled in, which was discovered restoration work, revealing the two apses situated alongside the so-called **cave of Santo Domingo**, the place imbued with great importance for the Order as it was where the founder would retire to pray and mortification; the cave was also subject to remodelling work in the 15th century, backing onto a chapel with a beautiful façade by Juan Guas and a sacristy. Its interior was prolifically decorated in line with the precepts of the Baroque; worthy of special note here is the sculpture of Santo Domingo de Guzmán carried out by **Sebastián de Almonacid**. It is currently the only outbuilding which still belongs to the Order as the rest of the monastery was expropriated after the Ecclesiastical confiscations of Mendizábal in the 19th century; it is now owned by the Provincial Government which has granted it on a temporary basis to IE University.

Coat-of-arms of the Catholic Monarchs

Of the monastery complex unit, the chapel **façade** has been preserved in very good condition which, by way of a retable, reveals an overlaying of different architectural structures: the basket arch, lintel, tympanum with lobulate arches and exterior archivolt broken to house a set of sculptures, whilst

ICONOGRAPHIC PROGRAMME

1. Saint Domingo de Guzmán
2. Saint Thomas of Aquinas
3. St. Peter of Verona
4. Saint Raimundo of Peñafort
5. Andrés Cabrera
6. King Fernando V, the Catholic
7. Joseph of Arimathea
8. Holy Virgin Mary
9. St. John the Evangelist
10. Isabel I, the Catholic
11. Beatriz de Bobadilla
12. Coats-of-arms of the Catholic Monarchs
13. Lope de Barrientos, the bishop of Segovia
14. Saint Vicente Ferrer
15. Christ on the cross
16. Angels with coats-of-arms of the Dominican order

the entablature and the pinnacles serve as an alfiz. The archivolts bear vegetal decoration and on the tympanum the theme of Pity can be observed, incorporating the presence of other characters in line with the thinking of the Catholic Monarchs regarding the participation of the people in the drama; hence, the scene includes the Virgin with the body of Christ in her lap, María Magdalene and Joseph of Arimathea; at the sides, the praying figures of the monarchs escorted by two figures who have been identified with Beatriz de Bobadilla and her husband Andrés Cabrera. The other theme represented is the Crucifixion which features two Dominica monks praying, identified as Saint Vicente Ferrer and Lope de Barrientos, the bishop of Segovia. Flanking this scene are two eagles with the royal coats-of-arms already featuring the pomegranate, indicating that the work is subsequent to the conquest of the city in 1492; at the corners of the upper part, the coats-of-arms belong to the Dominican order and they are supported by angels. The iconographic programme is completed by figures who appear at the sides under nitches and who represent four Dominican friars who have been identified as Saint Domingo de Guzmán, the founder of the Order, Saint Thomas of Aquinas, St. Peter of Verona and Saint Raimundo of Peñafort.

Chapel gateway

THE SEGOVIAN ROYAL MINT[79]

Entrance door

Wheels for hydraulic development

In Segovia, at least in its territory, coins have been minted since the time of the emperor Augustus whose effigy appears on a native minting where for the first time you can also read the word: SEGOVIA.

Shortly after the resettlement of the city whilst Alfonso VII was in power, the Emperor minted coins, though it was from the reign of Fernando IV onwards when regular minting started.

The «Old Mint» was located in the San Sebastian Corralillo which operated concurrently with the new one until 1681 when the last minting was carried out; minting was carried out there with a hammer.

In 1581 Felipe II asked his cousin Archduke Fernando II of Austria to buy a mint machine which the latter had set up at his mint of Hall in Tyrol, Austria. The machine, the most advanced of its time, allowed minting with a roller, using hydraulic power by means of several teethed wheels.

In 1583 the works on the Royal House began according to plans by Juan de Herrera who was finishing off the works at El Escorial, with the

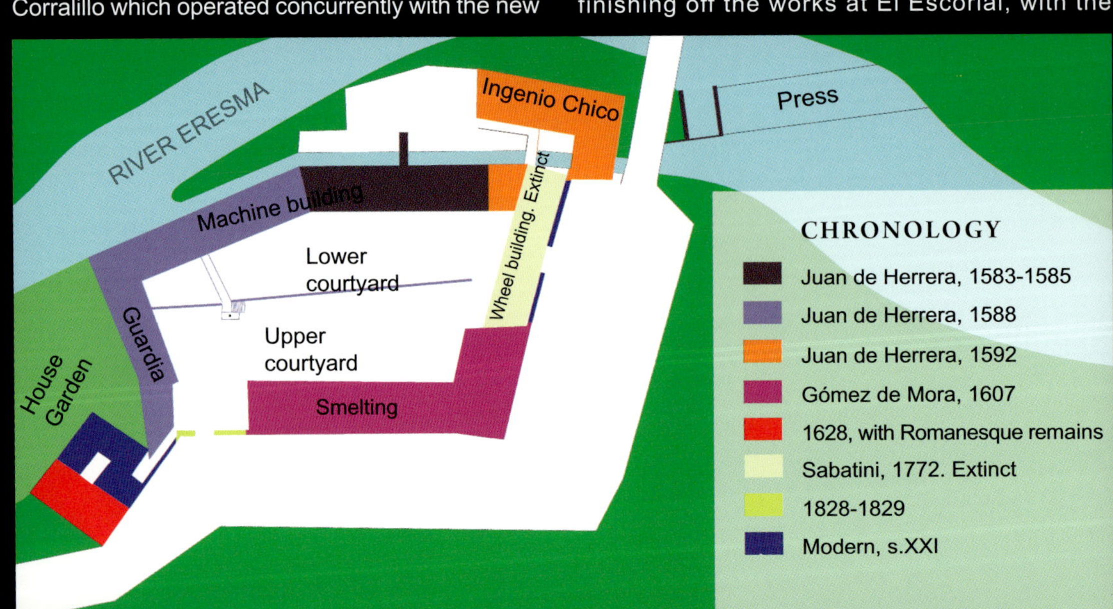

advice of Austrian technicians who had come to supervise the works. In 1583 the boxes arrived with two machines which were installed forthwith. Shortly afterwards, further mint machines were installed and the nave had to be expanded to get them to fit. In the same way, the containment wall was raised whose stairway replicates those of El Escorial. In 1592 the little machine was built intended for gold and silver coinage, with the large one being used for vellon and copper coins.

OPERATION

The Royal Mint worked like an autonomous city with its own laws and leisure zones, its own security corps and even dungeons, the workers lived at the factory itself on the upper floors of the naves. There was an aqueduct. For production by roll, firstly the alloy desired is melted into elongated bars (the rails) which are passed on successive occasions via various rollers until they have been sufficiently flattened. They were then whitewashed to clean the accumulated dirt and slag, again passing through two rollers which mark the relief of the coin. They are then taken to the cutting workshop where, striking them, they were separated from the die, now being ready for circulation. From 1719 onwards they started to apply the milled edge or drawing in the corner to avoid fraud by grinding. As from 1770 the roller minting system was dismantled and the presses by wheel system was introduced with the remodelling being undertaken by Sabatini, constructing a new wing and dismantling the surplus old machines, leaving only the rolling mills so that once the disks (coin blanks) had been struck, to them to the press by wheel and there engrave them and provide the milled edge; hence, the coins improved their minting quality, allowing them to be made more slender and stackable and, first and foremost, being able to include the effigy of the King which has appeared since that time. In 1865 automatic presses were installed though unfortunately in 1869 the provisional government decreed their closure to centralise all the production in Madrid, transferring all the recently installed machinery there.

Reconstruction of a device, Segovia Museum

From that point onwards the Royal House went through different hands which made it into a flour mill until in 1967 all activity stopped permanently. After years of ruin, fires and abandonment it was totally restored in 2011 after three years of works.

SOME COINAGE FROM THE ECSC OF SEGOVIA

Cincuentín, Felipe IV, Approx. 170 g ~ 70 mm
Silver ~ 50 REALS ~Actual size ~ Segovia Museum

8 Escudos, Felipe IV,
Approx. 27 g ~ 70 mm ~ Gold.
Actual size ~ Segovia Museum

2 Reals, Felipe V, Approx. 27 g ~ 26 mm ~ Silver. Actual size

4 Maravedís, Carlos IV,
Approx. 25mm ~ Copper.
Actual size

MONASTERIO DE SANTA MARÍA DE EL PARRAL[80]

The Parral from the Alcázar

HISTORY

The last Hieronymus Monastery was erected in Spain in what is known as Alameda de El Parral alongside the River Eresma, sheltered by a hillside which protects it from the cold of the north and supplies it with numerous fountains which convert its surroundings into an orchard as reflected by the Segovian expression: *From the Orchards to Parral, paradise on earth*, as the latter was an area chosen by numerous religious orders to establish their monasteries.

Its foundation is attributed to a legend according to which **Mr. Juan Pacheco, the Marquis of Villena**, in order to express his gratitude for having come out unscathed of a duel, promised to build at this place a monastery at the hermitage of Our Lady of Parral. Its construction was actually ordered by **Enrique IV** in **1447** though the Marquis of Villena took charge of the work, having built some humble constructions to shelter the monks during the monastery construction process. The first monks came from the monastery of Guadalupe and after a few years, without having undertaken any work, they decided to leave; it was at this time when King Enrique IV decided to take control of the foundation and start of the works, with the main chapel having fallen under his patronage which it is believed he was planning on using as a royal pantheon, though the concession of the funerary chapel finally ended up in the Marquis of Villena's house, his son, after the end of the civil war which set the heirs of Enrique IV against each other, made it hard to complete the works.

The monarchical unit we can now see was about to disappear in the early 20th century; in 1808 it was sacked by the French and in 1836 when their monastery was exclaustered due to Mendizábal's confiscation which is when the period of ruin, abandonment and sacking began, to such an extent that it was considered knocking it down owing to the poor condition of the constructions; however, in 1914 it was declared a National Monument and in 1925 the monks came back again and part of its lost splendour began to be recovered.

THE MONASTERY

First and foremost, worthy of special mention are certain aspects of the Hieronymus Order in order to understand the layout of the monastery. This is a Spanish Order which has never crossed the borders of the Iberian Peninsula, with its foundations being mainly in Spain and some in Portugal; they lived under the protection of the Crown which promoted them. The Order was founded in **1373** and its first monastery was St. Bartholomew's of Lupiana in Guadalajara, soon spreading to other places, highlighting the monasteries of Guadalupe and San Lorenzo de El Escorial.

In the early days they led a contemplative, hermetic life and then after their foundation they began to observe rules and a life as a community; this is why their foundations were always carried out on pre-existing hermitages.

The person responsible for carrying out the monastery design, following the precepts the latter should have, was **Juan Gallego** who put it at its current site in the place of a former hermitage and with four cloisters: for Reception, Hostelry, Chemist and the Main one or for the Processions.

The monastery began to be erected before the church in 1454 when King Enrique IV, in view of the idleness of the Marquis of Villena and the helplessness of the monks, decided to take control of the company; his patronage was borne out by royal coats-of-arms with pomegranates at various outbuildings.

As we have mentioned, it has four cloisters whereof, owing to closure, only that of **Reception**, of smaller dimensions, can be visited, with a portico built in the early 16th century serving as an antechamber with a tank and a garden from where you can behold beautiful views over the city, soothed by the sound of water from its fountain. From here you can see the ruins of the **Hostelry Cloister**, built by Enrique IV who had some lodgings here decorated by rich coffered ceilings; unfortunately, it suffered a fire in the mid-16th century. Several chapels open out onto the **Main Cloister or The Cloister of Processions**, endowed with large dimensions and worthy of special mention the 15th century richly decorated gateway on whose jambs there is a depiction of the theme of the Annunciation.

Gateway portico

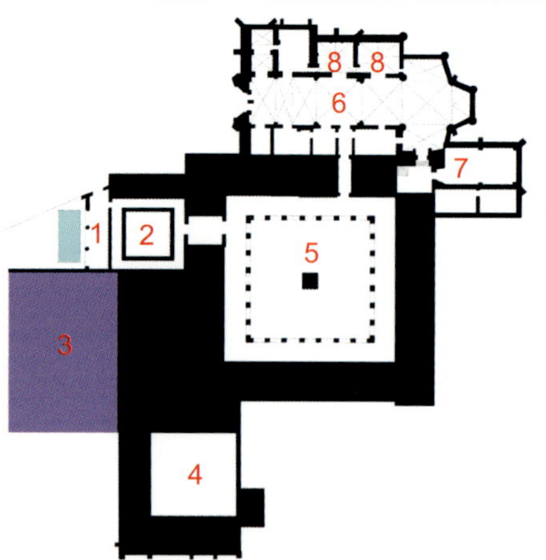

MONASTERY OF ST. MARY's OF EL PARRAL

1 *Gateway gallery*
2 *Gateway cloister*
3 *Hostelry cloister, ruins.*
4 *Chemist cloister*
5 *Main or Processions cloister*
6 *Church*
7 *Sacristy*
8 *Former late Romanesque hermitage*

CHURCH

The temple started to be built from the sanctuary; during the course of the works the hermitage continued to be used and it was finally included in the church, giving rise to two chapels: that of Sancho García del Espinar and Alonso González de la Hoz which were just one space at the beginning. It is assumed that this was the hermitage owing to the presence of late Romanesque eaves corresponding to remodelling work undertaken on the building during the 13th century.

As we have already mentioned, upon taking charge of the work Enrique IV decided to convert the main chapel into his place of burial, but it ended up as the family pantheon of the Marquis of Villena; the person responsible for finishing the works was his son Diego López Pacheco, the second Marquis of Villena. The church consist of just one nave with lateral chapels between buttresses connected to each other by narrow doors, the polygonal sanctuary with a unitary layout for the main chapel, the transept and the choir stall at the base, responding to the Hieronymus planimetry scheme. It was designed by **Juan Gallego** with the participation of **Juan Guas** and **Martín Sánchez Bonifacio**, responsible for closing the vaults in around 1485. The intervention of Juan Guas is evident on the central section of the sanctuary which is covered by ribbed arches comprising transept aches and tiercerons with the presence of ties from the Toledan school with drawings of arrow tips.

On the exterior the building looks incomplete as the **façade** has not been concluded. The gateway is endowed with structural and formal solutions relating it with the Monastery of Santa Cruz[122] and the cathedral cloister[50] in which **Juan Guas** took part as a drawer and **Sebastián de Almonacid** as a sculptor. In the centre the gateway opens out with two semicircular arches adorned with pomegranate branches alluding to Enrique IV and sheltered by a pointed arch framed by an alfiz. In the centre, the mullion, empty, sheltered under its canopy the image of the Virgin, currently located in the Cloister of the Processions. On the jambs there are two high-reliefs representing the Virgin with a phylactery and the angel in the scene of the Annunciation; they are in poor condition and without any heads. Flanking them are canopies endowed with rich tracery without any figures inside. In the middle section of the façade a large window opens out on which the heraldic coats-of-arms of Mr. Diego López Pacheco, the second Marquis of Villena, and of his wife Mrs. Juana Enríquez are framed.

The façade is completed by the silhouette of the bell tower, crowned by a richly carved cresting; it was crafted by Juan Campero, responsible for transferring the cloister from the old cathedral to the new one.

Pomegranates on the church door

Main chapel

ICONOGRAPHY OF THE MAIN CHAPEL OF THE

MAIN RETABLE

1 Life of St. Hieronymus
2 Coat-of-arms of the Hieronymus Order
3 Penitent St. Hieronymus
4 The Last Supper
5 The Wash Basin
6 The Virgin of Peace
7 The Birth of the Virgin
8 The Visitation
9 The Annunciation
10 The Hug before the Golden Door
11 The Circumcision
12 Pentecost
13 The Assumption of the Virgin

TOMB OF THE MARQUIS OF VILLENA

1 Hope
2 Adam
3 Justice
4 Temperance
5 Prudence
6 Eve
7 The Fortress
8 Mr. Juan Pacheco Marquis of Villena
9 The Three Mary's, St. John, Joseph of Arimathea and Nicodemus before the Body of Christ
10 Charity
11 Faith
12 St. Nicholas
13 St. Anthony
14 Angels with the Arma Christi
15 Coats-of-arms of the Marquis
16 Not identified
17 St. Stephen promartyr
18 St. Lawrence martyr
19 St.James the Moor Slayer

MONASTERY OF ST.MARY'S OF THE PARRAL

14 The Crucifixion
15 God the Father
16 Angels with the coats-of-arms of the House of Villena
17 The Prophets: Ezekiel, Isaiah, Daniel, Zachariah, Jeremiah and Malachi
18 Saints
19 St. Mark and St. John the Baptist
20 St. Hieronymus in his studies
21 Martyrdom of St. Andrew
22 Prophets
23 St. Matthew and St. Luke
24 The Baptism of Christ
25 St. Michael the Archangel
26 Prophets

TOMB OF THE MARQUISA OF VILLENA

1 Hope
2 Adam
3 Justice
4 Temperance
5 Prudence
6 Eve
7 The Fortress~Samsom
8 Mrs. María de Portocarrero, Marquise of Villena-
9 Holy Burial of Christ
10 Faith
11 Charity
12 Not identified
13 Angels with the Arma Christi
14 Coats-of-arms of the Marquisa
15 St. Helen
16 St. Lucia
17 The Appearance of Christ to his Mother alongside St. Pedro, St. James and St. John

CHURCH

Inside, at the base, the choir stall is raised on ribbed vaults which start off with angels bearing the coats-of-arms of the first and second Marquis of Villena and their respective wives. The choir loft opens out onto the nave by means of a beautiful basket arch with fringes.

As mentioned earlier, lateral chapels open out into the nave between which it is worth mentioning the **chapel of St. Hieronymus** which was the first to be built at the site of the old hermitage, being founded in 1482 by Mr. Alonso González de la Hoz, the secretary and accountant of Enrique IV, who was responsible for buying the hermitage and convinced the first monks not to return to the monastery of Guadalupe. It is followed by the Assumption chapel which, joined to the previous one, comprised the hermitage.

Another interesting chapel is that of the **Descent**, belonging to Fernán Pérez Coronel, previously Abraham Seneor[35], who converted to Christianity in 1492, having been christened at the Monastery of Guadalupe and whose godparents were the Catholic Monarchs.

The **main chapel** is lit up by six large windows flanked by the statues of the apostles undertaken by **Sebastián de Almonacid** and identified by their attributes and names on the halo; these sculptures are characterized by their clean lines and their static serenity. The coats-of-arms of the second marquis of Villena rest on them.

The **main retable** was carried out in line with the precepts of the early Renaissance; made from polychromed, golden wood, it consists of five aisles, four bodies and loft. The three central aisles are decorated by reliefs alluding to the life of the Virgin, the patron

Tomb of the Marquis of Villena

saint of the monastery; in the others saints and prophets are represented. Flanking the retable are the **tombs of Mr. Juan Pacheco and Mrs. María de Portocarrero** which were carved in alabaster in 1528 by Juan Rodríguez and Lucas Giraldo; both are endowed with a retable structure with the figures of the deceased in praying position inside a richly decorated niche whose bottom is decorated by the relief of the Three Maries before the tomb in the case of María and the Descent in the case of Juan. The two figures are attired in clothes of the time, with the marquis wearing armour decorated by grotesques and accompanied by a servant, whilst the marquise is wearing a bonnet and is accompanied by a lady. Around them, inside vaulted niches, there are a multitude of saints.

The chapel is closed off by a grating carried out in the early 17th century and having disappeared in the 19th century with the Ecclesiastical confiscations.

The **antesacristy doorway**, situated on the southern arm of the transept, entails a Baroquisation of the Hispano-Flemish structures seen up until now, there is no longer any basket arch, but rather a pointed arch whose exterior archivolt opens out to form a drawing that combines a hunched structure with another polylobed one. It is suspected that it was not a doorway originally, but rather it served as a frame for the adjoining **tomb** belonging to **Mrs. Beatriz de Pacheco**, the daughter of the first marquis of Villena. Its authorship is attributed to **Juan Guas** and **Egas Cueman**; in the heart-shaped structure there is a depiction of the enthronement of the Virgin and on the sides there are angels bearing the coats-of-arms of the marquis of Villena.

Antesacristy doorway

Sanctuary of St. Mark and the Alcázar

ST. MARK[82]

At the confluence of the Clamores stream with the River Eresma you will find this **12th century Romanesque** temple, one of the few examples that still survive of the numerous medieval parish churches that existed in this area. Thus is a simple building, with the temple having been transformed over time, but retaining its original physiognomy with just one nave and a semicircular apse with two gateways, surrounded by a wall configuring the space of the atrium. Inside the baptismal font from the Romanesque era prevails.

THE SEGOVIAN ROOFTOPS

It is worth pointing out the curious appearance of the rooftops which are apparently missing the upper tiles, only having those of the canal; this can be put down to the fact that with this layout the snow is prevented from accumulating and loading the excess tiles.

ST. BLAISE[81]

Near the previous church there lies the remains of the **Romanesque** church of San Blas, built in the late **12th century** and converted into a private house today; thanks to this the sanctuary has been preserved with its **semicircular apse** and an interesting window and a gateway. This church, along with the aforementioned St. Mark's and the now extinct ones of St. James and San Gil, constituted an important settlement nucleus in the Middle Ages which, over time, has gradually disappeared as have their parish churches.

Sanctuary of St. Blaise

BAREFOOT CARMELITES CONVENT [84]

Tomb of St. John of the Cross

This spot on the edge of the River Eresma was chosen by **St. John of the Cross** to found his convent. To carry out his project, the saint was aided by Mrs. Ana de Mercado y Peñalosa whose tomb is preserved in a chapel of the church; it is said that St. John of the Cross himself worked on the convent construction works.

The convent church seems sombre and simple on the exterior, following the Carmelite line in the **17th century**; by contrast the interior, endowed with just one nave, is decorated by plasterwork frames which lend it a richer appearance. In one of its side chapels there is the **tomb** of the saint richly decorated with reliefs which recount episodes from his life and Carmelite saint figures.

Convent Entrance

ST.JOHN OF THE CROSS

He was born in Fontiveros in the province of Ávila in 1542, going by the name of Juan de Yepes Álvarez. In his youth he joined the Carmelite Order which he left after meeting **St. Teresa of Jesus** whom he joined to undertake the Carmelite Reform, resulting in the new Order or Barefoot Carmelites. Along with St. Teresa, and by himself too, he went around part of the Peninsula founding new convents; in this way he reached Segovia in 1574 with St. Teresa to found the St. Joseph's Convent[53]. He returned to the city years later to take charge of the Barefoot Carmelites convent which he had founded himself. During the course of his life he was not only devoted to the Reformation, he also undertook important literary activity which earned him the title of patron saint of Spanish poets and his proclamation as a **Doctor of the Church** in 1926 by Pope Pius XI. He died in late 1591 in Úbeda and his remains were transferred to his convent in Segovia where they lie in a luxurious tomb made by Félix Granda in 1926.

SANCTUARY OF THE VIRGIN OF FUENCISLA[85]

Under the Peñas Grajeras, at the confluence of the Clamores stream with the River Eresma, there lies the temple that house the patron saint of Segovia. Her names stems from the Latin expression **fons stillans**, a dripping fountain or a fountain that drips, in reference to the vast amount of springs and fountains to be found in the area. It seems that since the Middle Ages some construction already existed in the area which was extended at a later date and still proved insufficient, having been replaced by the current Baroque one erected in the late **16th century**, following the designs set out by **Francisco de Mora** and under the supervision of **Pedro de Brizuela**.

The building is simple, both in terms of its appearance and ground plan, highlighting its beautiful interior where the image of the Virgin is kept; this is a late medieval carving which holds the Baby Jesus in its hand with the stateliness characteristic of his style. It is now a vestment image, though it was not made for this purpose as the rear part is not carved. Its feast day is celebrated in September and many legends have arisen around the figure, some were included in the **Poems of Alfonso X**.

Fuencisla Gate

Alongside the sanctuary, on the highway to Arévalo, this commemorative arch appears, a work from the early 18th century, designed by Juan de Ferreras alongside the Madrid Gate. It is a totally Baroque work, richly decorated and highlighting the relief which makes reference to the miracle of María del Salto.

MARÍA DEL SALTO

The best known legend related with the Virgin of Fuencisla is the one which tells the story of the Jewish Ester; this woman was sentenced by Jewish judges to be thrown from the heights of Peñas Grajeras for having committed adultery as she intended to marry a Christian man. Before she was thrown to her death, the woman begged the Virgin to intervene on her behalf and when she was hurled into the void a miracle occurred as she hit the ground softly and without being hurt in any way. The woman was christened Mary and given the nickname la del Salto.

VERA CRUZ[83]

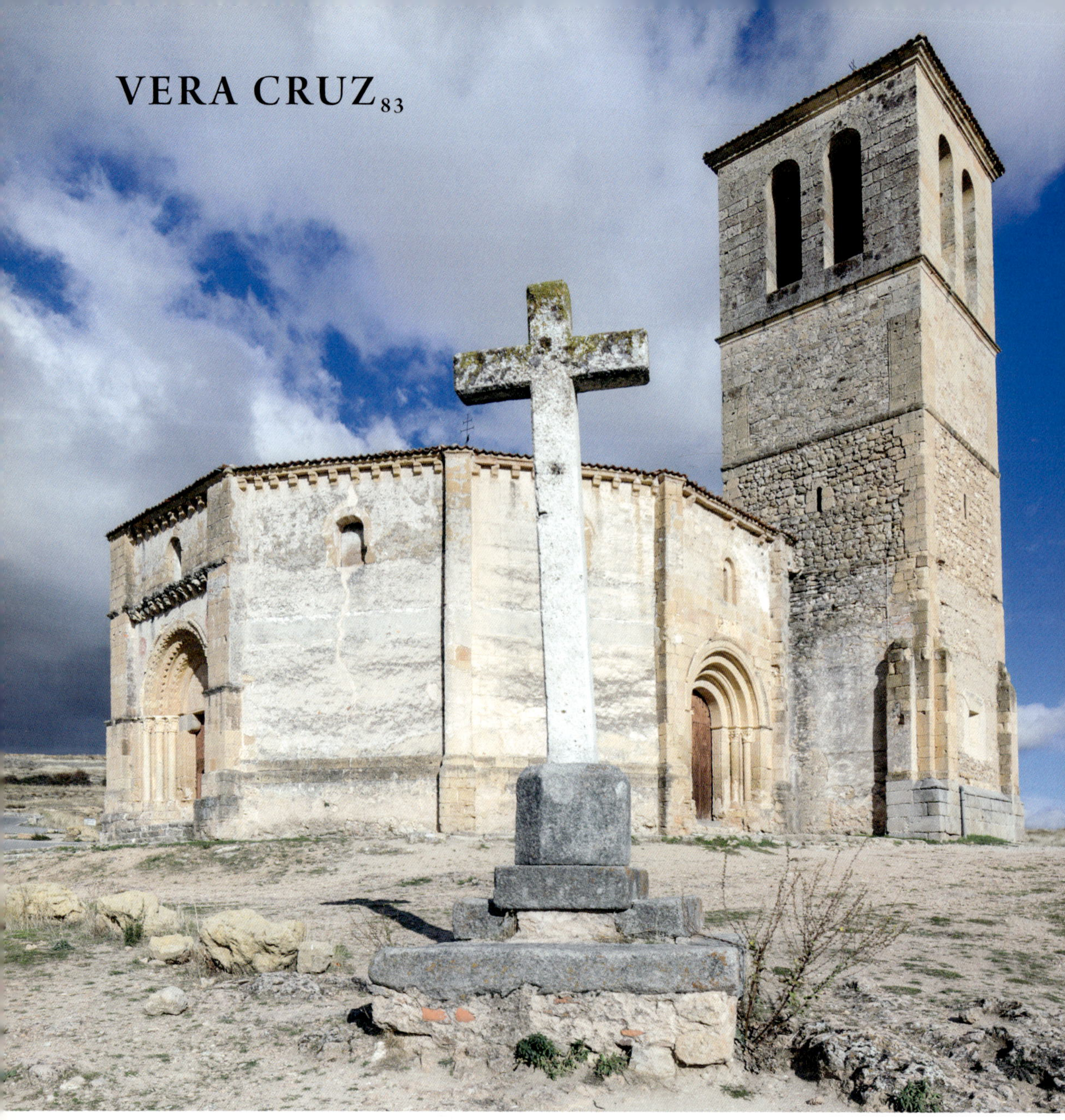

The foundation of this late Romanesque church, built in the first half of the **13th century**, has traditionally been attributed to the Knights Templar, but in reality it can be put down to the **Order of the Holy Sepulchre**, currently owned by the Order of St. John of Jerusalem to whom it was transferred after the union of both.

Generally speaking, the churches of the Military Orders had a centralised layout and this type was related during Antiquity with commemorative actions, then moving onto the Paleochristian world and being reflected in the anastasis of Jerusalem; they could have circular layouts, like S.Mark's church in Salamanca, or polygonal, as is the case in point and the churches of Torres del Río and Eunate in Navarre, with the latter being endowed with funereal connotations.

The church of Vera Cruz is endowed with complex morphology and a **centralised** dodecagonal layout with two concentric bodies and a sanctuary with three semicircular apses; the most important part of the whole is the central structure which is surrounded by a deambulatory covered by a barrel vault and which opens out to the sanctuary. In this building it has been endeavoured to combine the concept of martirium and church.

The central squinch comprises two overlaid bodies; the inferior is a dodecagonal space covered by a lowered dome supported on short shaft columns owing to the elevation of the ground and four accesses oriented towards the four points of the compass. The upper body is closed by a ribbed vault whose ribs do not cross in the central space as is also the case at the church of San Milan[114], being an Islamic-based structure. The purpose of this upper body is not known, though there is an altar table there whose front is decorated by intertwined arches which recall one of the corridors of the cloister of San Juan de Duero in Soria.

Another unknown aspect at this temple is the existence of a tiny room in the northern part whose purpose is also unknown and there is a theory that it could have been an access to a disappeared lantern, others believe that it could have been a kind of cell for penitents or the place where the archive or the Treasury were kept, though its liturgical use cannot be ruled out either; there are more rooms which can only be accessed from the exterior and which are smaller.

This temple seems to wish to imitate the Church of the Resurrection in Jerusalem, presenting certain similarities with the church of Tomar in Portugal.

In the lower body of the tower you will find the **Lignum Crucis** chapel, donated in the early 16th century by the Marquise of Denia; the relic is currently at the parish church of the neighbouring town of Zamarramala.

On the exterior the building gives away its interior structure; it has two gateways, the western one with zigzag decoration at the archivolts and historiated capitals, whilst the southern one is simpler and has a built-in late Romanesque relief which is highly eroded depicting the scene of the Three Marys before the sepulchre of Christ.

REAL SITIO DE SAN ILDEFONSO

SAN ILDEFONSO ROYAL PALACE

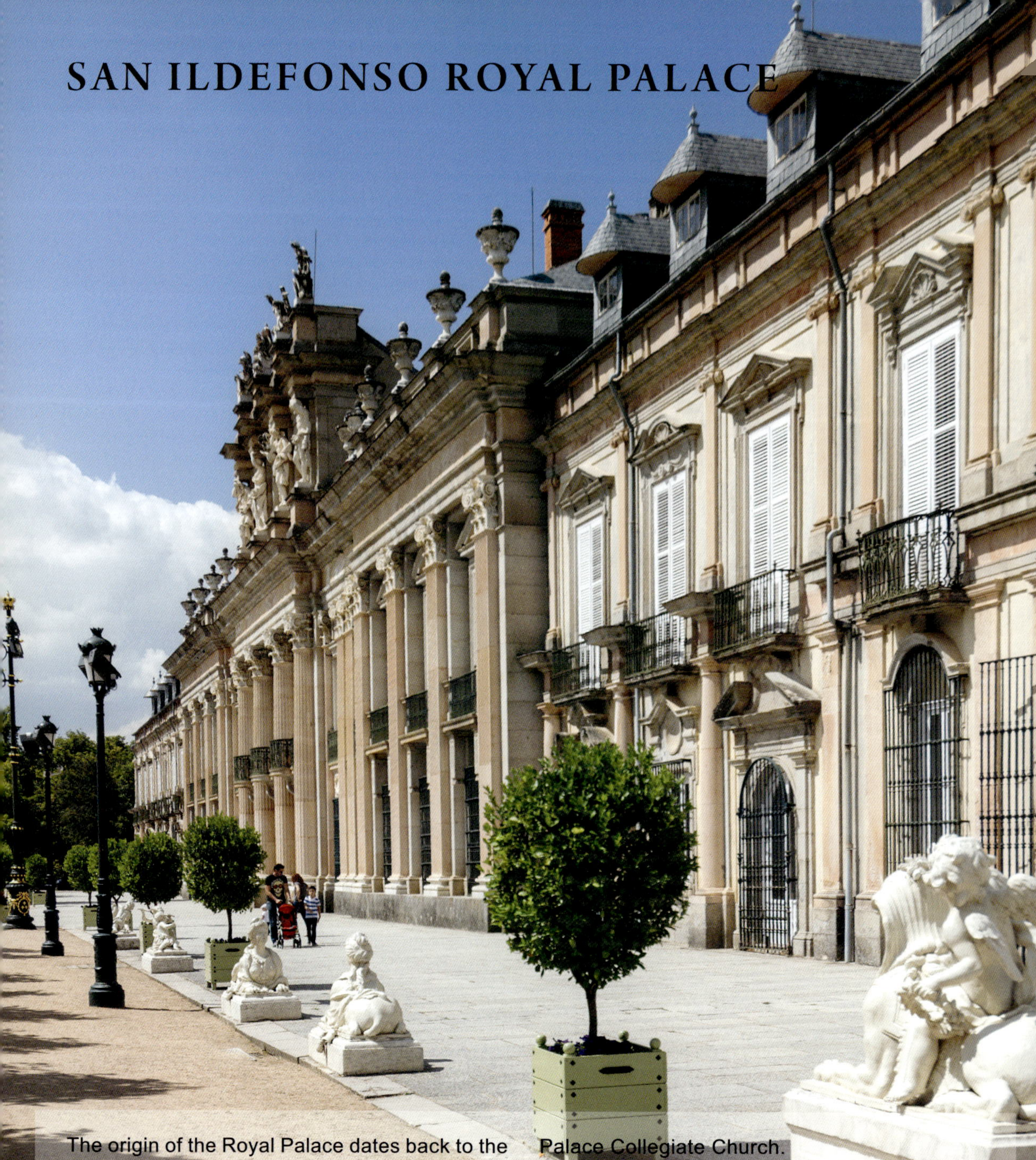

The origin of the Royal Palace dates back to the farm that the Hieronymus monks of El Parral[130] had at this spot; it was here that there was originally a hermitage dedicated to St. Ildefonso which Enrique IV had built and which the Catholic Monarchs later donated to the monks of the Segovian monastery. **Felipe V** decided to build a new palace in this area after the ruin and abandonment of the nearby palace of Valsaín, becoming his fixed abode for more than 20 years; what's more, both he and his second wife **Isabel de Farnesio** are at rest in the Royal Palace Collegiate Church.

The palace, built in Versailles-style with clear Italian influences, was the summer residence of the monarchs of Spain until Alfonso XIII. The architects involved in the design were Teodoro de Ardemans, Procaccini, Felipe Juvara and Sachetti. The palace interiors are richly decorated according to the tastes of the 18th and 19th centuries with marbles and murals al fresco as well as furniture and impressive lamps from the Royal Glass Factory[150] located just a few metres away.

The Horse Race

La Selva

Diana's Baths

In this palace complex the most noteworthy aspect are the gardens which surround the palace and which provide a perfect example of 18th century gardens; its alleys, parterres and the layout of the fountains were designed by a team of engineers, architects and sculptors, resulting in some beautiful gardens which in some areas have been converted into an authentic forest. Between the trees planted pride of place goes to the limes, horse chestnuts, cedars and the vast secoyas. The gardens are splattered with statues of mythological characters and extravagant urns made of white marble. However, what this palace and its gardens are well known for are the fountains, made of lead and painted so as to imitate bronze, with mythological themes in some of which the water jet reaches 40 metres high as is the case of the Fama fountain. All the fountains are supplied by a large pool known as the "Sea" which is situated in the highest area of the gardens, managing to supply all the fountains owing to the strength of the differences in level. There are many fountains but the most beautiful, impressive ones are the Horse Race, the Anfitride Waterfall, the Diana Baths, the Fama or the Canastillo.

Detail of the Cascada Nueva

ROYAL GLASS FACTORY OF LA GRANJA
NATIONAL GLASS CENTRE FOUNDATION

The creation of the Royal Glass Factory must be linked to the remodelling work undertake in the **18th century** by the Borbons, the new reigning dynasty; during this period, particularly in the Illustration, and until the reign of Isabel II, the formation of industries was pursued which were sufficiently strong and innovative so that it would not be necessary to resort to the importing of luxury items from abroad. With this in mind, the Borbons saw to the supply of as much latest technology machinery as well as specialized and foreign teachers to develop and teach the subjects as in Spain there was no qualified labour.

At the beginning there were three factories within the Real Sitio location, but after a fire the new one was created outside the palace enclosure. The Royal Factory started to be built in **1770** following the drawings by **Joseph Díaz, alias Gamones**, a builder from Real Sitio; it had a basilica-shaped layout with two transepts covered by refractory brick vaults under which the furnaces were located. At the end of the century it had to be expanded by Juan de Villanueva. Its façade is sober, merely adorned by the coat-of-arms of the Borbons.

It experienced a period of great splendour, largely owing to the manufacture of flat windows or mirrors, but after the death of Fernando VII in 1833 production was abandoned and it was rented out

Furnace

Smoothing

Mould blowing

One of the pieces manufactured

to private individuals, being permanently abandoned in 1972. In 1982 the **National Glass Centre Foundation** was formed with a view to recovering not only the factory, but also production and the recovery of techniques, creating a **Glazing School** and a **Research Centre** as well as a **Museum** which displays pieces that have been carried out traditionally at the factory alongside others of a contemporary nature, thereby announcing the evolution of glass craftsmanship; the old production machinery was also recovered and restored.

Nave with furnaces and machinery, museum

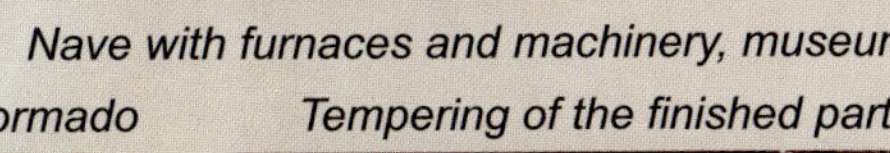

Sticking of parts

Formado

Tempering of the finished parts

ROYAL PALACE OF RIOFRÍO

The construction of this Real Sitio was commissioned in **1751** by the widow queen **Isabel de Farnesio** during the reign of her stepson Fernando VI, as her personal abode; when her son Carlos III came to the throne, the palace took a back seat and was converted into a **hunting pavilion**. It was only inhabited on a temporary basis by the consort Francisco de Asís, the husband of Isabel II and by Alfonso XII who settled there after the death of his wife María de las Mercedes. The building is set in a 625-hectare forest inhabited by a great number of **bucks and stags** which has made it be housed in **Hunting Museum**. It was designed by the Italian architect Virgilio Rabaglio in 1751, denoting an austere, appearance to the exterior, contrasting with its richly decorated interior, being notable for its main stairway which is divided into two, contributing the showmanship characteristic of the Baroque; it houses important works of art.

VALSAÍN

The mountains of Valsaín, on the northern side of the **Guadarrama Mountains**, have always been the source of wealth, both in terms of cynegetic development and the exploitation of its natural resources; at present, its mountains are exploited on a sustainable basis by the wood industry thanks to the **wild pines** which originate from its land.

As early as the Middle Ages it was reserved for hunting by the Castilian monarchs; hence, Enrique III ordered the construction of what is known as Casa del Bosque (Forest House) which his grandson, **Enrique IV**, expanded and turned into a hunting pavilion in view of his predilection for Segovia and this activity, having, in the past, accommodated exotic animals there like lions and an elephant. However, it was with **Felipe II** that it became a true Flemish-style **palace**; it was used almost continuously until the reign of Carlos II during whose reign it suffered a fire which burned it down and was not restored as it was decided to build the nearby palace of San Ildefonso during the reign of Felipe V, deploying some elements of the previous one, which exacerbated its state of ruin and final abandonment. Of all that splendour, where one of the most noteworthy rooms was the **Mirrors Room**, only a dilapidated tower remains standing and the arches carried out by Gómez de Mora; the place for important festivals such as those experienced on the grounds of the wedding of Felipe II and Ana de Austria or the birth of Princess Isabel Clara Eugenia, the daughter of the latter and Isabel de Valois, have withstood the test of time.

THE PROVINCE OF SEGOVIA

Gorges of the Duratón and hermitage of San Frutos

SEPÚLVEDA

This medieval town, declared as a **Historical-Artistic Unit** in 1951, is located on a hill overlooking the River Duratón; its existence dates back to the presence of some kind of prehistoric settlement, but until the 10th century it was not really developed; it appears documented for the first time in the *Chronicles of Alfonso III*, referring to the reign of Alfonso I who made some incursions this way. It was resettled in 940 by Count **Fernán González**; as tradition would have it, the city was taken after a confrontation between the count and a Saracen, resulting in the victory of the former who cut off the Moor's head which, according to the legend, is to be found on the façade of Casa de los Proaño or of the Moor.

The territory of Sepúlveda belonged to the Crown of Aragon for while under the reign of Alfonso I the Battler as part of the conflict between the latter and his wife Queen Urraca of Castile.

The medieval town preserves part of the 10th century **city wall** and its gates, in actual fact Sepúlveda is known as the **Town of the Seven Gates**. It also has numerous Romanesque churches, with the most noteworthy being that devoted to the **The Saviour**, dating back to the late 11th century and regarded as the oldest Romanesque church in the province.

GORGES OF THE DURATÓN

The odd landscape formed by the broad windings of the River Duratón accommodates one of the most important colonies of **griffon vultures** in Europe with more than 500 members which use the rocky escarpments at around **100 metres** high for nesting. Thanks to its rich landscapes, fauna and flora, it was declared a Natural Park in 1989. In addition to its rich natural heritage, it also conceals a rich artistic heritage, highlighting the prehistoric caves and the **hermitage of San Frutos** situated in a peninsula in which, so legend has it, the Segovian saint retired to pray in the 7th century in the Company of his brothers St. Valentine and St. Engracia; the church and the remains of the buildings preserved today belong to the former Romanesque monastery of Benedictines founded there during the 12th century. Access can be gained via a bridge which crosses a Deep gorge, called Cuchillada de San Frutos, attributed, according to tradition, to the saint who drove his stick into the ground, forming the gorge to protect the residents from the Moslem area.

THE ROMANESQUE IN THE PROVINCE OF SEGOVIA

The province of Segovia boasts one of the densest and richest Romanesque heritages, highlighting the churches perched along the valley of the river Duratón and those built at the foot of the mountains. Worthy of special mention are the sets of **Sacramenia** and **Fuentidueña** which, although their heritage has been reduced owing to the transfer of their churches to the United States, still retain a good example of the best Romanesque temples and the porticoed galleries are of special note whose capitals have created a school.

A very important building within the Segovian Romanesque is the church of St.Michael of **Sotosalbos**, near the capital, which, surrounded by a well looked-after environment, boasts a porticoed gallery with a very diverse repertoire of themes on its capitals.

However, the jewel in the crown of the provincial Romanesque is Duratón. This town is located alongside the stream of the river of the same name; in its vicinity there lies the **archaeological site** of the old Roman city of **Confluentia** which Ptolomeo talked about. This city was also inhabited during the Visigoth period and the important **burial site** dates from this time whose trousseaus are preserved and are on display at the Segovia Museum[6]; we could say that the archaeological site at Duratón is the most important one in the whole province. In its vicinity you will find the **Romanesque church** of St.Mary of the Assumption whose sculptures have served as the model for many temples, both in the province and in the capital which have been inspired by the **capitals of the portico**, even copying its formal structure; these include the historiated capitals devoted to Christ's childhood and that dedicated to the **Birth** and the **Epiphany** is very noteworthy. These capitals – as well as the corbels and metopes - are characterised by their delicate, refined carving and their wealth of detail.

Fuentidueña

Sotosalbos

Duratón

PEDRAZA

This **medieval town** has been able to preserve its essence admirably; after crossing is only access gate in the city wall, we are surprised by an urban fabric which has stood still in time which is why the town was declared a Historic Site in 1951 and has served as the backdrop and stage for many cinema, TV and advertising shoots and its paved **Main Square** is very well known worldwide. The height of its splendour coincided with the height of the wool trade in the 16th century and the majority of its **houses and palaces** derive from this period. In addition to its city walls it boasts a **castle** erected in the 13th century and remodelled in subsequent centuries; it was acquired in 1926 by the painter **Ignacio Zuloaga** who restored it to locate his workshop there; between its walls it currently houses a museum dedicated to the artist. Other notable monuments in the town are the Prison and the Romanesque church of St. John's with its narrow tower.

TURÉGANO

The settlement of the **castle** on the hill dates back to the Celtiberian era and since that time its occupation must have been continuous until the erection of the fortress. Both the Town and the castle were donated in the 12th century to the Bishopric of Segovia, which is why the bishop in office Arias Dávila promoted the remodelling of the fortress in the 15th century. Inside its walls there lies the **Romanesque church of St. Michael's** and it is an enigma which of them was built first, the castle or the church. For a time the fortress as used as a prison and one of its most renowned inmates was **Antonio Pérez**, the secretary to Felipe II.

CUÉLLAR

The area of Cuéllar was resettled in the late 11th century, though there was a prior attempt by the Counts of Monzón during the 10th century which was frustrated by an attack by Almanzor; however, its time of greatest splendour occurred in the 12th and 13th centuries at which time its **city walls** and castle were raised as well as all its **churches** in **Romanesque-Mudejar** style, thus being regarded as the **"Mudejar town"**. In the mid-15th century the town and its castle were handed over by Enrique IV to his royal favourite **Beltrán de la Cueva**, the Duke of Alburquerque.

In addition to its rich heritage, in its history the town of Cuéllar has the oldest bull runs in Spain.

SANTA MARÍA LA REAL DE NIEVA

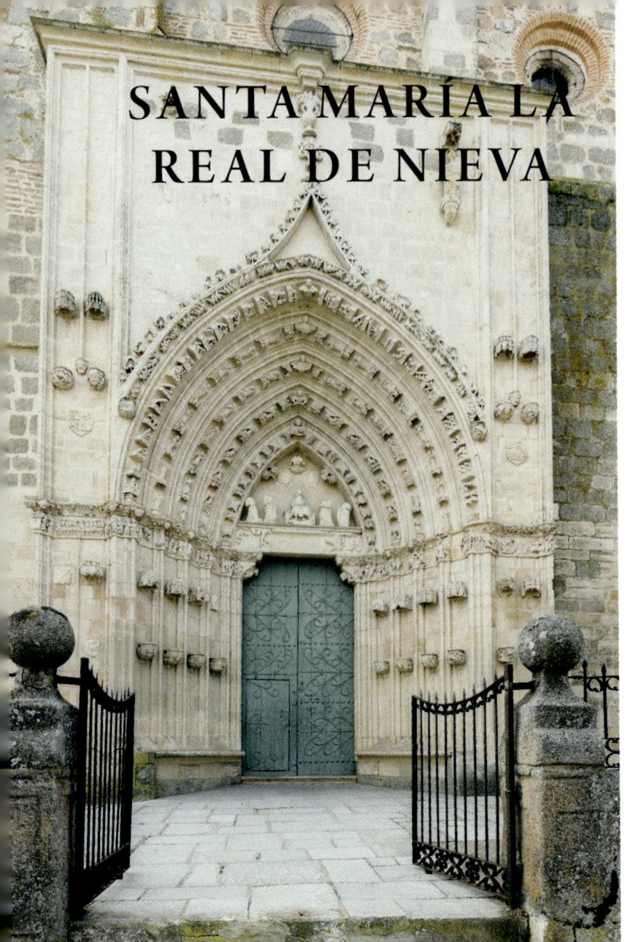

This town was founded in the late **14th century** under the patronage of Queen **Catherine of Lancaster** after the discovery of the image of the Virgin in a slate quarry where the church is situated today; the image, known as **Our Lady of Soterraña**, was found by a shepherd in 1392 and the construction of the hermitage was ordered thereof which was soon to become the church we can see today and a monastery under the aegis of the Dominicans. Houses began to be built around the monastery, giving rise to the town whose settlers were exempt from taxation and from going on military campaigns. The original image of the Virgin was greatly damaged in the fire which occurred in the early 20th century and the current work is by Aniceto Marinas, keeping in its interior the remains of the old one.

At the monastery, in **Gothic** style, worthy of special note are the **cloister**, endowed with late Romanesque connotations, and a wide variety of themes on its capitals, and the **north gate** on which there is depiction of the Final Judgement. For several centuries queen **Blanca I of Navarre** - who died suddenly in the town - was buried here.

COCA ~ CAVCA

Since antiquity this has been a unique human settlement as it is situated at the confluence of the rivers Voltoya and Eresma; in their documents the Romans already talked about the Vaccean city of **Cauca,** highly populated and which in the 2nd century was to become a Roman municipality. During this time it was an extremely important place in view of its archaeological findings in the city itself and on its outskirts; hence, from the Pre-Roman era three verraco sculptures have been preserved, one of them ingrained on the castle walls and from the Roman times the remains of old buildings and villas have been found. Its

importance was on a such a scale that the Roman emperor **Theodosius the Great** was born here. The most noteworthy monument in the town is the **castle** whose construction was commissioned in the second half of the **15th century** by Alonso de **Fonseca**, the Lord of Coca, whose family is buried at St.Mary's Church in tombs made of Carrara marble by Doménico Fancelli and Bartolomé Ordóñez. The castle is done out in **Gothic-Mudejar** style, made of brick, whose interplay of volumes lends great plasticity to the whole.

ALPHABETIC INDEX

ALBERTO FERNANDEZ FERRERO ET LAURA ILLANA GUTIERREZ ME FECERUNT

IN DIE SCT. GENADII ASTVRICENSIS DIES MERCVRII VIII KAL. IVNII MMXVI A.D. HOC LIBRUM IMPRESSUM EST IN BURGIS